WHAT OTHERS ARE SAYING...

All private pilots are trained to follow a sequence of actions with an engine out emergency. It's freezing cold, you probably can't make the airport, your daughter is on board, and absolutely every second counts. Would you and your co-pilot do everything exactly right, and just what exactly would right be? Steve Jackson, my flight instructor years later, recounts in breath taking detail the events and actions taken by himself and co-pilot from engine out to crashing in the Pennsylvania forest. A touch of faith is present too that provides a *never-say-die* approach to the problem solving even as the plane goes into the trees. Every pilot should read this account and consider it part of their training for emergencies.

—Joseph M. Zias,
Private Pilot,
retired Professional Engineer
and Assistant Professor.

As the author's daughter and one of the survivors that walked away from this plane crash, I can tell you how hard it is to put into words the emotions and thoughts that raced through my mind that day and how I came to terms with the possibility of death. However, my father's narrative of our crash perfectly describes the fear and trauma that we faced, and reading this made me feel like I was reliving the most fearful moments of my life. The author has a way of pulling you into his mind and feeling his every emotion, and he also portrays the story of a man's faith and how the bond between a father and daughter became strengthened by such uncertainty and doom. I highly recommend for those who want to feel inspired.

—Chandi Jo Brooks,
Author's Daughter & Survivor

N1XF IS GOING DOWN

Steve Jackson

N1XF IS GOING DOWN

We're Alive Thank God We're Alive

TATE PUBLISHING
AND ENTERPRISES, LLC

Published by Tate Publishing & Enterprises, LLC
127 E. Trade Center Terrace | Mustang, Oklahoma 73064 USA
1.888.361.9473 | www.tatepublishing.com

Tate Publishing is committed to excellence in the publishing industry. The company reflects the philosophy established by the founders, based on Psalm 68:11,
"The Lord gave the word and great was the company of those who published it."

Cover design by Joseph Emnace
Interior design by Gram Telen
Photo by Jay K. Bradish—Bradford Era, Bradford, PA

Published in the United States of America

ISBN: 978-1-62994-473-9
1. Biography & Autobiography / Personal Memoirs
2. Biography & Autobiography / Adventurers & Explorers
14.01.23

DEDICATION

I dedicate this book to the memory of my mother, Maryetta Jackson, who taught me what unselfish love is and lived her life totally committed to the furthering of God's kingdom.

ACKNOWLEDGMENT

To my wife, Joan, my daughter, Chandi, and my son, Chad, your love and support has propelled me to success and enjoyment unattainable without you. You have allowed me to fly high and often.

CONTENTS

IT'S A GREAT DAY TO FLY

Chandi, Pat, and I were trudging through the knee-deep snow. Pat was hollering, "We're alive. Thank God we are all alive." Chandi was on her cell phone trying to reach her friend Burke, but there wasn't a signal. I was trying to lift my short legs high enough to walk through the deep snow to find my way to the logging road about thirty yards ahead, not sure if I should sport a wide smile or fight back tears because God had carried the three of us through this terrible accident with only a small scratch from the shattered windshield on each of us. For Chandi and me, we had just been part of an event that was sure to strengthen our already strong father-and-daughter relationship. The engine failure of N1XF on this cold January afternoon put us in the trees short of Runway 23 at the Bradford Regional Airport.

January days in the northeast United States can be cold. Real cold! January 10, 2004, was a typical winter day in western New York State as I awoke from my deep sleep.

This was the kind of day pilots dream about. I expected to have a clear day since a high pressure system was prevalent in the northeast. The winds were to be light and variable, and since the Lycoming 260 horsepower engine in the Comanche is air cooled, this would be an ideal day for maximum performance. In addition, I anticipated the air to be stable, which would make for a smooth ride today.

The day was bitter cold with over two feet of snow on the ground. It had been weeks since the first frost, the ground was rock hard, and the snow was several inches deep. You could feel the firmness of the earth as you walked across the hard-packed icy snow track. The temperature was well below freezing, about one degree Fahrenheit below zero.

As I jumped out of bed this morning, the first thought that raced across my mind was what a beautiful day it is to fly. I glanced out my window to see the coldness of the three-foot-high snow piles as the wind blew the loose snow across my driveway. There wasn't much wind, but enough to make the top of the snowbanks disappear in the white blanket swirling above them. I heard the hot water baseboard heat start to crackle as

I turned the thermostat up and headed to the shower to get ready for my flight to Syracuse, New York, from Bradford, Pennsylvania, and then back to Bradford.

On my way to the shower room, I stopped to look out the hallway window and looked up to see a beautiful, clear, blue sky. I noticed the white streak across the sky over Olean, New York, that gorgeous winter morning, one put there by a high-flying jet on a Southerly heading, most likely leaving the area on a flight originating from Buffalo. As I so often do, I glazed at the magnificence of a huge jet, jetting through the air at a high altitude. Almost every day, I would look up and ponder what life would have been had I became a pilot at a much earlier age.

As I often do, I recalled the first time the thought of a career as a professional pilot crossed my mind. I was standing in the living room of the home I grew up in with my new rifle in my hands. I really wasn't interested in hunting, but all my buddies were hunters so I had the idea I should be also. I had the stock of the rifle against my arm looking through the high-powered scope attached to the top of my gun when I heard my mother say, "Dean is going to school to be a pilot." Dean is a year, maybe two years, younger than I am and lived across the road from us. I immediately dropped the gun to my side then laid it on the chair beside me and said, "Mom, what did you just say?"

She repeated, "Dean is going to school to be a pilot," and I almost instantly felt a little jealousy even though I had never considered piloting as a career. From that moment on and for the next twenty-five years, I envisioned myself piloting an aircraft in some capacity. It was common for me to stare into the air when a plane would fly by and say to my wife, "I really wished I had become a pilot."

In the fall of 1999, my son Chad, my wife, Joni, and I got in our car; and I drove to Findlay, Ohio, to visit our daughter. While in Findlay, I drove by a small community airport, the Findlay City Airport, and once again marveled at the Cessna 172 flying around in the pattern practicing takeoff and landings. At that time, my wife said, "If that is what you really want to do, why don't you just go do it?"

"Because I cannot afford to fly as often as I want to, and I would want to at least become a flight instructor, if not an airline pilot," I replied. Joni responded by saying, "Then we will sell our business, you will go learn how to fly, and you can work toward your dream of flying." She continued, "We have raised our children, and now it's just us. It's time to do what we want to do." Oh my goodness! How many spouses would be so supportive and be willing to so drastically change our lives as we approached the age of forty?

At the time, I was the sole owner and franchisee for an H&R Block office in Eldred. Our busy season was quickly approaching, so I wasn't able to act on my wife's offer immediately. So when I returned home, I focused my efforts toward preparing for our busy season. Most of my staff routinely returned and so did my husband-and-wife team of Jill and Shane. I shared my dreams with Jill and Shane one day while we were on break from a training session and was surprised to learn of their interest in purchasing the franchise from me. I soon entered negotiations with them and quickly forged a fair agreement with them. I assisted Jill and Shane with securing the financing and was pleased when everything apparently fell into place as my prospective buyers were promptly approved.

The deal came apart one day when Shane informed me that he and his wife had reconsidered and weren't sure they were ready for such an undertaking. By this time, I had already been accepted into Spartan College of Aeronautics in Tulsa, Oklahoma, and had been aggressively marketing my home in Portville, New York. I am sure the disappointment was evident in my face and in the tone of my voice as Shane and I discussed the uncertainty he and Jill were experiencing.

After a long and disappointing day at the office, I was ready to go home and tell my wife the news that we had to put our plans on hold. I slumped into the driver's seat of my car and cranked up the radio to lose myself in my thoughts. As soon as I turned onto Main Street, the local news was blaring across my radio. The headline that evening was that Jamestown Community College in Olean and Jamestown had today announced they intended to implement an associate of science program in professional piloting. I am a man that has a deep faith in God, but this was really crazy! I know he is in control except this was the most amazing timing I have ever experienced. It validated my goal and at the same time changed my plans. I couldn't wait to get started and called JCC as soon as I thought they were open the following day.

I was able to speak to Rick, the gentleman responsible for managing the professional piloting program at JCC that morning, and when he told me they were still waiting for New York State approval, I ignored him and informed him I would be the first to enroll, just tell me what to do! Rick replied he would keep in touch with me and offer me that opportunity.

After several more conversations with Rick, we developed a plan to get me started before the fall semester when JCC would be rolling out their new program. The plan was to get started with Gary at Wellsville Flying Service in Wellsville, New York, and JCC would allow me credit for the hours of training I received prior to beginning my formal education at JCC. This could be allowed because the actual flight training was to be contracted out from JCC, and successful completion of each course was when a student passed their check ride.

I immediately started twice-a-week lessons in Wellsville, and when the time arrived to start at JCC, I was almost ready to complete my initial course and become a private pilot. I worked hard to stay on schedule and graduated from JCC in the summer of 2002 with my associate of science in professional piloting.

I had continued to operate the income tax preparation office while attending school, which was a combination that became very manageable through the help of my office manager, Ginny. In addition to my responsibilities at my business, I started to work part-time at Jamestown Aviation as a flight instructor. The flight instructors endorsement was for me the most difficult to achieve. It also was the final course needed to earn my degree, and it is generally the key piece for aspiring professional pilots. It is the endorsement that allows a pilot to log hours in large enough numbers that can be credited toward required hours for experience purposes, which open up career opportunities in the field of professional piloting.

My personality is such that I was soon investigating avenues to work as a freelance instructor. I made an attempt to set up in Olean, New York, and when that didn't materialize, I was welcomed at the Bradford Regional Airport outside Bradford, Pennsylvania.

It was at Bradford that I met Rob who became a student of mine. While a student, Rob purchased the Piper Comanche that I would be flying today. He also owned a few different manufacturing plants, one which was located in Maine and the reason for this flight today, and the flight I intended to make the following Monday.

It seemed like only seconds but had been minutes that I stood there peering out that window lost in the thought of piloting an airplane across the sky, something I knew I would be doing within a few hours.

I broke out of my trance when I heard my wife yell from the bedroom that I needed to get myself in motion or I would be late. So I turned and went in the shower room to freshen up and prepare myself for the day ahead, not realizing how this January day would change my life forever. On this day, I felt my inner self to be anxious to fly, more so than on most days. I was again deep in thought going over the details of our flight for such a long period of time that I finally realized the shower water was no longer hot, not even warm but ice cold. It was time to exit the shower and rush around as I slipped on my trousers, shirt, and other clothing with my mind still filled with excitement over this wonderful winter day that would help the Piper Comanche I would be piloting perform amazing well.

I leaned over and kissed my wife good-bye as I was ready to hurry down the stairs to my computer so I could sneak one last peak at the weather forecast and review my flight plan. As my wife, Joni, regularly did, she told me to have fun, stay safe, and that she loved me. I stopped to stare at her and remind her I was always cautious. I usually responded by reminding her how safe flying is, but on this day, I turned and said "Yea, OK" and hurried out of the room.

I sat down at my desk, reached down to power on my computer and happened to once again glance out of the window; this time, I had a different perspective. I was on the first floor now and wasn't able to look out over the neighbor's house to see the swirling snow. *It didn't look quite so cold from this view*, I thought as a smile crossed my face. The ice on the side of the house had a stunning reflection in it that radiated the warmth of the sun starting to peak up over the Allegany Mountains. I was so deep in thought that when the computer beeped as the windows program started to open, it startled me.

Once at my desktop, I quickly logged into my account at DUATS, or Direct User Access Terminal Service, a service provided to the aviation community for the purpose of obtaining weather briefings, scheduling, and filing flight plans with air traffic control and other services. As I waited for the home page to load, I opened my IFR low altitude charts to review my course of flight, frequencies, and approach charts that provide all the detail necessary to execute instrument approved approaches and landings at airports. I reviewed my DUATS-generated flight plan, altitudes, on course headings, anticipated frequencies, and projected ground speeds. Officially, DUATS, or Direct User Access Terminal Service is a weather information and flight plan processing service contracted by the Federal Aviation Administration (FAA) for use by United States civil pilots and other authorized users. I also noted how many practice approaches and which ones I intended to fly at the Syracuse Airport.

An IFR flight is one that is flown under instrument flight rules. These rules are different than when flying with visual references; although on this day, I could easily fly visually since there is unlimited visibility and it is a perfectly safe day to fly. Most experienced pilots like to fly under IFR flight rules, even on good flying days like today since it keeps us in direct and constant contact with air traffic control (ATC). IFR does require advanced training and an additional endorsement. The great advantage to having this endorsement is there are many days, like the one I expect next Monday when I fly to Maine that we can fly when we couldn't under visual flight rules (VFR).

The final task I needed to complete before opening the front door and braving the elements was to call Patrick. Pat was to be my safety pilot for the day. A safety pilot was not required for this IFR flight, but Pat had been one of my trusted instructors as I was going through the various stages of my training. I trust Pat and admire his flying abilities. He had been in Afghanistan flying not as a pilot but as part of the crew. In addition to possessing many hours of flight instruction as an instructor, Pat had extensive experience flying in a Piper Navajo as a captain for a cargo company. A cargo pilot generally will fly in difficult flying conditions, and this experience makes most cargo pilots good IFR pilots.The purpose of this trip was to sharpen my instrument flying skills for a flight I expected to captain on the following Monday to Maine. The area weather forecast for this Monday was inclement weather and solid instrument flying conditions. I couldn't think of any friend or fellow pilot I felt more comfortable flying with for this purpose than my associate Pat.

I dug into my pocket to locate my cell phone that stored Pat's number. As I flipped open my phone, I caught a glimpse of the snow blowing past my kitchen window and I again thought what a beautiful day to be in the air. Once at altitude, I was sure it would be a smooth flight, and I could picture seeing from the air the hills of western New York and northwestern Pennsylvania with the frost and snow clinging to the branches.

I found Pat's number and quickly dialed it on my land phone, and after three rings, I heard Pat's deep voice say, "Good morning, Steve. What a great day we have to fly."

I said, "Hey, Pat, I am just ready to leave my house for the airport. Is there anything you need for the flight?"

He responded with a "Heck no! Let's just get there and enjoy this awesome day God has given us to fly." Pat lived in the Buffalo area so he had a long drive to the Bradford Regional Airport near Bradford Pennsylvania, about one hundred miles from his home and about a two-hour drive.

Wintertime in western New York and the Buffalo region is well documented. It gets extremely cold at times in January in Buffalo, and this day wasn't an exception. On extremely cold days, ice pellets are visible in the air, and at a temperature reading below zero on this Saturday morning, one could see the cold. Add to the below-zero temperature, the wind, which was only about six knots, and you get to feel what is called the wind-chill factor. The wind-chill factor is what the temperature feels like in the wind and is usually several degrees colder than the actual temperature.

On this day, the temperature at the Bradford Regional Airport was even colder than in Buffalo. The high temperature for Buffalo for January 10, 2004, was twenty-six degrees Fahrenheit, and in Bradford, the high for the day was only nineteen degrees Fahrenheit.

To drive from Buffalo to Bradford, Pat would be traveling Route 219 South from just south of Buffalo into Pennsylvania, through the city of Bradford until he would turn onto Route 59 where the Bradford Regional Airport is located. It is a pretty drive through the hills starting with what is referred to the south towns of Buffalo. Orchard Park, Hamburg, Boston, Springville, and into Ellicottville, which is known as "ski country" throughout the northeast United States.

The south towns are where the snowbelt area is. When one thinks of Buffalo, they not only think of cold winter days and nights but of huge snowstorms like the storm that is referred to as the "Blizzard of 1977". The "Blizzard of 1977" was a deadly blizzard that hit the region from January 28 to February 1, 1977.That week, the wind gusts were form forty-six to forty-nine miles per hour, blowing snow onto the highways, making it near impossible to clear the roadways do to the wind packing the snow so thick that the snow became solid.

On this day, it wasn't nearly as bad as the sky was a beautiful blue without a cloud visible anywhere. However, the beauty of traveling the windy path of Route 219 through the snowcapped mountains made the trip for Pat pass by quickly. As pilots, we are always anxious to get to the airport and generally manage to visualize our flight as we travel the highway to get there. Pat had two hours, and I had an hour on the road.

After a few minutes on the phone with Pat reviewing our flight plan and confirming what approaches I wanted to practice, it was time to pack up my charts and flight plan and get out to my car. Pat was just pulling out of his driveway, so I knew I had some extra time but I was anxious to get to the airport.

I stuffed my charts into my flight bag, shut down my computer, and grabbed a banana on my way out of the kitchen and hollered upstairs to my wife that I was leaving. As I opened the front door to walk onto the porch for my short walk to the garage, I was immediately aware of how cold this day was in Olean, New York. A couple of deep breaths later, I stepped outside on to the porch and heard the crack of walking on a frozen wooden deck. I certainly wasn't going to sneak up on anyone this day. One could hear the noise of car engines idling in the cold as my neighbors were trying to get some heat in their vehicles before heading out onto the road. In my excitement of planning my trip and thinking about piloting the 260-horsepower Piper, I hadn't even considered warming up the car. Within minutes, my hair was stiff from the cold, and I thought, *Oh, boy, the first ten minutes of this ride will be terribly cold.*

My garage didn't have a door on it, but my car was under cover, so frost on the windshield wasn't an issue. I would only need to let the engine warm up for a short time and didn't need to scrape any windows or wait for the defroster to clear the windshield. So I opened the trunk and threw my flight bag in, shut the trunk, and jumped into the driver seat of my car.

When it is near zero degrees, I always wonder if my car will start. I am old enough to remember the days of carburetors on cars instead of fuel injection systems. But with the fuel injection systems on cars today, cold weather usually doesn't mean a hard start, so as I turned the key, the engine quickly fired and I was ready for my road trip to Bradford Regional Airport.

A SLOW TRIP TO THE AIRPORT

I backed my car out of the garage, and it is almost like my car sensed the cold also. Most January days in western New York are so cold that the condensation in the exhaust system always makes the car smoke. On this morning, the exhaust smoke was so thick it was just about like driving through dense fog.

As I put the car in drive and headed out of the driveway, I saw my neighbor Craig walking to his car. I unrolled my window and exchanged well wishes with him. Craig asked where I was headed on this cold day, and I proudly told him I was going to practice some instrument approaches at the Syracuse Airport, and he seemed somewhat surprised. His comment was "Oh my God, it is way too cold to fly today."

I sort of chuckled and said, "It is never too cold to fly, fuel doesn't freeze, and we have heat in the airplane."

As Craig walked away, he turned and said, "I don't know about that, but be careful."

"OK, Craig," I replied. "Have a good day, and I'll see you later."

Next to my home in Olean was a Country Fair convenience store that was open twenty-four hours and was always busy. I am not a coffee drinker or a smoker, but I do have a bad habit; I am a huge Pepsi drinker and like to drink one in the morning. So I turned left onto West State Street and into the Country Fair to buy my morning caffeine. I had some time to waste before Pat would arrive at the airport, so I bought a *Buffalo News* newspaper and made sure I read the sports pages while I allowed the interior of my car to get to a comfortable temperature.

There are three different routes I often take to get to the Bradford Airport. Most times, I would drive east on Route 417 through the village of Portville and then turn south onto Route 305 into Pennsylvania, which would take me through the towns of Eldred, Smethport, and up to the top of the hill named Mt. Alton where the airport is. On other occasions, I would take Route 16 south through Knapps Creek and down the Pennsylvania side of the hill into Derrick City, then up another hill through Cyclone and to Mt. Alton. This route was the most scenic since the road was a fairly straight one to the top of the Knapp Creek hill but would wind between the trees on the Pennsylvania side of the hill. And after turning in Derrick City, it would then wind up through the forest to the top of another hill. Once on top of the hill, the road would follow the top ridge across to Mt. Alton. On a cold day like this, it is such a stunning sight to drive through the forest on these secondary roads. Often I would see different wildlife or at the very least evidence that I had missed them by only minutes.

The route I chose to travel on this day is relevant to the details of my story, which I will share with you as my day unfolds. I decided to take Route 16 south to the Barnum Road, which also winds through the woods near the top of Rock City Hill. Once on the Barnum Road, I would drive down the twisted road leading down the hill and, within a few miles, reach the Pennsylvania border. Once the road enters Pennsylvania, it takes you across a low area we refer to as the flats in our area. The Indian Creek flats near Eldred are a road separated by a swamp area on each side. My wife doesn't like driving this road as she often has worried about losing control of her car and ending up buried in this swamp with all the reptiles. On this morning, as I drove through the flats and noticed the ice build up, being buried in the swamp fleetingly crossed my mind. I once again looked toward the sky and thought even worse I hope I never fly an airplane into a swamp. It would plow so deep I would never get out.

As I approached Devil's Elbow, an intersection with four corners with a dangerous dip as you approach the intersection from the east (the reason for the name),

I needed to make a decision on which way to go. If I went straight onto the West Eldred Road and followed the river on the West side, I could pick up Route 446, which would, in about five miles, connect to Route 446 and then follow Route 446 to Smethport; or I could make a left turn and cross the Kendall Bridge where I could connect with Route 446, which is the main street in Eldred. After passing through Eldred, I would drive about another six miles and pass the intersection that connects the West Eldred Road and Route 446.

I stopped at the intersection, unsure of which route I wanted to take on this morning. As I instinctively turned left forward the Kendall Bridge, I thought I have plenty of time; I think I will stop at the office and see what the girls were working at today. If I chose this way, I also could stop at Chandi's and invite her to go flying with me.

The office is the H&R Block income tax preparation office on Main Street in Eldred. I have been involved with this office and in this industry for many years as a franchisee and also as an employee of the local franchisee. The business has been part of my family since the 1960s, and we are well established in the area. However, January 10 was a day we probably wouldn't be serving many clients since most taxpayers didn't have their filing forms yet. As part of corporate policy, it was mandatory that we were open on this day, even though we didn't expect to be very busy. Besides, there were several last-minute tasks that needed to be attended to so we were well staffed for the day.

I had elected not to work this day so I would have the time to sharpen my instrument flying skills for the anticipated trip on the following Monday. My office manager, Ginny, and our receptionist, Dar, were working along with our new hire, Linda. I neatly parallel parked my car on Main Street, which I usually don't do, but I did on this morning.

There is a wooden walkway alongside the insurance building that leads to our office door. I jogged under the street sign that overhangs the walkway and quickly got to the door that opens into our office reception area.

The first person to greet me was Ginny. Ginny is now a franchisee, but on this day, she was the office manager. She also has always been a good friend of our family and one of my best friends. Ginny also is a person that worries about everything, and on this morning, I was grilled about why I was flying, where I was flying to, and how long I expected to be in the air. She didn't mind that I wasn't around the office on this Saturday; in fact, she probably appreciated that I wasn't there. Sort of like an overprotective mother that was always concerned for my safety.

As we jokingly chided back and forth, it became real clear that Ginny didn't think it was safe to fly on a day like this. Pilots know that cold clear days do not produce any additional risk than other good visual days, so I finally reminded her she always worried too much. I finished our lighthearted conversation by telling her I was going to stop at Chandi's apartment and invite her to go along with us. This was a spontaneous thought from me when I turned left at the Devil's Elbow, and I decided to follow through with it. It also accomplished my goal by sending her off, muttering something about being glad she isn't from Mars like men are.

Chandi is my only daughter and has supported me with this endeavor at every critical juncture. She was the one who would always send me a card as a check ride was approaching, offering her unwavering support. As I passed each check ride within days, I could expect another card from her, reminding me how proud she was that I was pursuing my dream. I hope she understands how important those short and simple messages were to me.

Chandi was in her midtwenties and rented a small apartment from my mother and father. The apartment was attached to my parents' home and located about two miles south of Eldred on Route 446. I would be passing right by her place on my way to the airport, and she had numerous times indicated she would love to fly with me if I ever had an empty seat and when I was flying out of the local area.

As I started my car and pulled back onto Main Street in Eldred, I decided for certain I would stop and pick her up if she were available to go with us this day. So as I approached her place and turned into the driveway that she shared with my parents, I became more excited that I might share this day with her. Out of the car I jumped, leaving it running to keep it warm inside, and I knocked on her door, wondering if she was even up yet.

The door opened, and she was surprised to see me.

"Come on in, Dad" she said. "Are you on your way to the airport?"

"Yes," I replied.

"I was wondering if you would like to go with me today."

"Right now?" she questioned. Then she uttered the standard comments: "My hair isn't done yet, and I am not dressed to go out yet."

"Okay," I said. "If you would like to go just be at the airport in an hour and fifteen minutes." After a few minutes of small talk, I was ready to finally go to the airport when Chandi said, "I would love to go. I will be there in a little over an hour. But if I am not there, just leave without me." I left uncertain if she would meet me at the airport or not.

I had only been at Chandi's about ten minutes, but now I figured I was on a tight time schedule and should arrive at the airport about the time Pat arrived. As I approached the intersection of Route 446 and Route 6 entering the town of Smethport, about twelve miles from the airport, I glanced to my left looking for traffic and noticed the local Ace Hardware store was open.

Great, I thought, *I should stop and get a pair of gloves for today.* So I shot straight across the intersection and drove into the parking lot and parked directly in front of the main entrance into the store. I don't wear gloves ever! But for a reason, I didn't yet know I felt the need to have a pair with me on our flight. I drove the remainder of the way, taking about twenty minutes before pulling into the service road of the Bradford Regional Airport. I turned right onto the service road that curls through the trees to the upper parking lot, which was on the left. I had agreed to meet Pat at this lot, so I turned into it and pointed the front of the car toward the service road entrance where I waited for Pat to arrive.

PREPARING FOR FLIGHT

As I sat in my car waiting for Pat's automobile to appear on the service road, I began to go over the steps necessary to prepare for takeoff. I once again pulled out my flight plan to review anticipated times and notably the amount of fuel required for this trip. By my calculations, I would burn about twenty-eight gallons of fuel and would need at least the required thirty minutes of reserve according to regulation. Daytime flights are required to add this reserve to the anticipated fuel burn as a precaution for miscalculation or the need to fly to an alternate airport. I routinely have a personal minimum of one-hour reserve, but on this day, I would carry well over three hours reserve with me.

After about five minutes, I saw Pat's car appear as he drove down the service road leading to the upper parking lot that I was parked in. Pat pulled alongside me and unrolled his window wearing a huge smile. "This is absolutely a beautiful day, Steve," Pat said, "and I cannot wait to get in the air."

"I am sure glad you're here, Pat" I replied. "I have already called flight service for the briefing, and it should be a smooth ride to Syracuse and back. The weather in Syracuse is about the same as here, so let's go to the hangar and get the preflight inspection done, just follow me to the hangar and we can park alongside it."

I slowly pulled away so Pat would have time to turn his car around and follow me onto the service road leading to the gated and secure area of the airport. I pulled up to the gate, pushed the button in my car to lower my window so I could scan my security card at the gate. Once the gate opened, I pulled through to the secure area with Pat following close behind in his car. The hangar where the Comanche is stored is on the back side of the second row of hangars, so I needed to drive along the community hangar, make a right turn, and turn onto the snow-covered blacktop area leading to the hangar. We then parked our cars at the far end of the hangar area so our automobiles would be far removed from any area that might be needed for arriving or departing aircraft.

Pat and I both exited our vehicles, and as we were walking to the hangar, I told him that my daughter,

Chandi, would probably be joining us for the flight. Pat replied that he had met my son since Chad had been a student for over a year at the flight school in Dunkirk where Pat and I first met, but he hadn't met Chandi yet and was anxious to have the opportunity.

The hangar doors are a combination of an electric bifold door that opens up in an envelop style and a walk-in door cut into the overhead door. So as we approached the hangar, I unlocked the walk-in door and stepped into the hangar with Pat behind me. The hangar doesn't have much extra room in it, especially with an airplane the size of the Comanche in it, so I quickly walked over to the light switch so we wouldn't trip or walk into something without expecting it. Adjacent to the light switch is the handle to raise and lower the overhead door so I pushed up on the lever to engage the wench system that would open the hangar door for us.

As the wench tightened, the wire cables to lift the door open, one could hear the loud snap of the cold cables and sheet metal door. As the door opened further, the light of the day and glare of the sun reflecting off the snow brightened the hangar to the point that the electric lights were no longer a help, so I walked over and turned the interior hangar lights off.

The Piper Comanche that we were flying this day is a good-looking bird. Tan with darker brown stripes and the letters N1XF on both the left and right side of the fuselage. N1XF is what is generally referred to as the "N" number or tail number. It serves the same purpose as a license plate on an automobile except that the tail number will usually follow the aircraft from one owner to another and won't be transferred from one aircraft to another aircraft. On a cold day, like this day, we would need to apply some heat to the under portion of the engine for about fifteen minutes to heat the metal of the engine, which would warm and thin the oil inside the crankcase to the point it could easily circulate and lubricate all the moving parts as needed. So I moved the portable heater into place, pointing the lower end of the long and narrow heater toward the underside of the front cowling.

No pilot may operate any aircraft without first verifying that all systems and equipment listed on a minimum equipment list are properly operating. Federal Aviation Regulation 14 CFR Part 91.205 "(b) provides a complete list of all items required on any aircraft flying daytime under Visual flight rules, (c) provides additional equipment required for night rules and (d) provides information for instrument flight rules which we will be flying under for this day." The balance of any minimum equipment required is provided with the minimum equipment list of each specific aircraft. Each airplane has a complete checklist that must be followed entirely to verify the airworthiness of the airplane for each flight.

While the heater was applying much-needed heat to the engine compartment Pat and I quickly walked around N1XF, verifying all surfaces were free from damage and verifying that the rudder, aileron, and flap controls were secured properly and appeared to be in good working order. A huge dent, ding, or rip in any surface could drastically alter the flight characteristics of the airplane or even make the airplane impossible to control in the air. It is also proper preflight inspection to visually inspect hinges and other connectors on all movable controls. This is the exterior portion of the preflight inspection required before boarding the aircraft.

Upon verification that the exterior portion of the N1XF was in airworthy condition, I shut off the heater and removed it from the proximity of the plane while Pat attached the manual tow bar to the nosewheel. Pat put one hand on the propeller while I positioned myself to the trailing edge of the left wing and began to push with Pat pulling to roll N1XF out of the hangar and into position to taxi.

While Pat disconnected the tow bar and returned it to its proper place in the hangar, I climbed onto the wing and slid into the cabin where I turned on the aviation master switch and dialed in the correct frequency to call the Bradford fuel office. Many airports have self-service pumps, but at the Bradford Regional Airport, they still provide fuel service at the hangar by truck delivery systems. I requested fuel service and verified that the position lights, beacon, landing lights, and strobes were all operating properly. I then shut off the avionics and electrical system and climbed back out onto the wing and then stepped to the ground where Pat was waiting with the checklist in hand. The next item was to verify tire pressure, oil quantity, a visual check of the break system, and verifying that all air inlets were not obstructed.

Before the fuel truck arrived, we needed to visually confirm fuel quantities in each tank. N1XF was equipped with six fuel tanks. A tank on each wingtip with a fifteen-gallon capacity in each tank, outer tanks on each wing with a thirty-gallon capacity on each side and the main tanks, with the same capacity as the outer tanks. The wing tip tanks each were about one half full for a total of fifteen gallons, the outer tanks were each showing approximately one half full also, totaling about thirty gallons. The inner tanks would be the main tanks we would use for this flight and I would have the lineman fill these to capacity, which would give us a total of about 105 gallons on board for this flight. I had calculated total fuel burn for this flight to be approximately twenty-eight gallons so we would launch with far more fuel than we would need to be legal and safe for this trip. Our legal requirement was anticipated fuel burn plus thirty minutes of reserve for a daytime flight. Adding the thirty minutes of reserve, we would need to launch with at least thirty-three gallons on board.

About the time Pat and I completed the necessary checks of these systems, I heard the clank of the diesel engine on the fuel truck as it approached the airplane. Brian hopped out of the truck, and as he rolled out the cable to ground the airplane, I instructed him to top off (fill) the inner tanks for me. He carefully removed the cap to the right tank and inserted the fuel nozzle as he filled the tank and replaced the cap. As Brian moved to the left main tank, I removed the cap and visually confirmed the right main tank to be full of fuel.

As Brian rolled the fuel hose bank to the fuel truck and disconnected the ground cable, I did the same to the left main tank, again verifying the tank was filled to capacity.

As was usually the case, Brian was interested in what our plans were for the day. Pat and I chatted with him for several minutes as we danced around trying to stay warm, but we needed to allow a few minutes for the fuel to settle in the main tanks. It takes several minutes for any containment such as dirt or water to settle to the bottom of the tanks. Once at the bottom, we could sump each tank and check for any noticeable problems with the fuel quality. After a few minutes, it was time to verify the integrity of our fuel, which we did at each tank looking for dirt or water. Flakes of dirt would be easy to see and any visible water would be noticed by what looks like bubbles since water and fuel won't mix, and with water being heavier than fuel, the water would settle to the lowest part of the tank where the sumps are located. The only thing left to do was to verify the quality of the fuel in the lines, which was done by placing a container beneath the sump located on the underbelly of the airplane, pulling a lever located on the floor of the cabin by the fuel selector, and then looking at the discharged fuel in the container. Once this task was complete, both Pat and I were satisfied that N1XF was indeed airworthy for today's flight.

While chatting with Brian, my cell phone rang, and it was Chandi calling from her car at the security gate. After completing our checklist, I jumped back into my car and drove over to the security gate to allow my daughter to enter and follow me back to the area where Pat and I were parking our cars.

Now we were ready to board the aircraft for what we thought would be our leisure flight to Syracuse and back to Bradford. I was going to be the pilot in command, the pilot who was responsible for the flight and in charge of any decisions that might be required while in flight, so I was in the left seat. I moved the right seat forward and made room for Chandi to climb into the back seat. I turned around and helped her fasten the seatbelt snuggly around her then slid the seat back as far as it would go so Pat could get into the right front seat where he could act as my safety pilot for the flight.

Pat buckled himself in, locked his door, and handed me the checklist so I could continue the interior preflight checks. The normal procedure is to brief the passengers on what to expect and what not to do, confirm between the pilots who is the pilot in command and which pilot will be handling the airplane, and to determine how we would communicate handing off any of the responsibilities in the event of an emergency. Then the mixture would be moved to full rich, the throttle would be pumped three to four times to prime the engine, I then would holler out the side window "prop clear," which would alert anyone in the area I am about to start the airplane. I, then, will look around the aircraft to make sure it is safe to engage the starter. Once started, I will need to verify oil pressure and fuel pressures, turn off the electric fuel pump, and check the fuel pressure again before moving down the checklist to monitor the local ASOS (automated surface observation system) on frequency 133.82 for temperature, barometric pressure, wind direction, and wind strength. The last item on the checklist before my taxi call is to verify transmission and receiving of my primary radio frequency of 123.07. A typical call goes like this: Bradford Unicom, November One X-ray Foxtrot (phonetic alphabet for N1XF) radio check please. Unicom would come back with a call: November One X-ray Foxtrot loud and clear. I then would respond, One X-ray Foxtrot.

So now the preflight is complete, and we are ready for our taxi call to alert all traffic on the ground, and in the air, about where we are and what our intentions and movements will be.

FINAL PREPARATION FOR FLIGHT

"Bradford traffic, November One X-ray Foxtrot taxing from the hangar, crossing the ramp to taxiway Alpha, holding short of Runway Three-Two on Alpha" was my call as I reached forward with my right hand, wrapping my fingers around the black handle as I pushed forward, opening the throttle to start the airplane moving forward. I watched as the RPMs went to about eleven hundred and the manifold pressure started to rise. As I pulled the yoke back toward me to take the weight off the nosewheel, the brown and tan Piper Comanche I was about to pilot started to roll forward. Once the plane started to roll, I very slightly brought the power lever back, lowering the RPMs to about one thousand to slow the aircraft to a fast-paced walk or slow-jog pace.

The taxiways and ramp leading to Runway 32 were all plowed but snow covered on this day. Along the sides of the taxiways, the snow was piled high, and the prop wash was blowing the loose snow around to the point it looked like a mini blizzard outside the airplane. "What a beautiful sight," Chandi said, adding, "I am really glad you asked me to come along, Dad."

"We're going to have great visibility today, Chandi," Pat interjected. "Just sit back and enjoy like I am going to while your dad works his tail off today."

"What work, Pat? This isn't work at all," I said. "But you two can relax since I am the pilot in command," I continued. "And, Pat, while you are relaxing, you can watch for air traffic for me since that's why I asked you to come with us today. If you don't want to, I will roll a wing up and dump you out," I joked.

To steer the airplane on the ground, the rudder foot petals are the primary controls. If you want the plane to turn left, you push on the left rudder pedal, and to go right, you would push on the right rudder pedal. The foot pedals are split with the bottom half being the rudder and the top half of the pedals being toe breaks that are differential, meaning the left pedal works the left main brake and the right toe break pedal will control the break on the right main tire. It can be tricky to slow the airplane while getting even pressure on the brakes so the airplane doesn't pull either right or left.

As I approached the taxiway leading to the ramp, I needed to turn left so I pushed the left rudder pedal as I pulled back on the yoke. The reason for pulling back on the yoke is it is a way of taking some of the weight off the nosewheel. The nosewheel and the rudder are connected with cables and work together so less weight on the nose tire will make the rudder pedals operate easier.

The taxiway leading to the ramp area leads us past the hangar that houses the shop area of our mechanic, the IA, which stands for inspector aircraft. Bruce had inspected N1XF several times and actually was the broker who sold the airplane to Rob. Bruce and I had flown N1XF on a few occasions, the latest being to Burke Lake Front (KBKL) near Cleveland, Ohio, just a week earlier. Bruce flagged us down, and we stopped for a quick chat. He hadn't met Pat before, so through the small side window that falls open, I introduced Pat to Bruce and told Bruce what our intentions were for the day. We chatted about what a great day we had to fly, and Bruce asked if there was anything we needed help with, which we replied no. As is common, he simply said, "Have a safe flight," and he would chat with us when we returned to Bradford.

Once again, I moved the throttle forward and started to taxi N1XF to the ramp area. There are three taxiways at the Bradford area, neither of which is parallel to the runways so we approached the closest one to the threshold for Runway 32, which is taxiway Alpha (A). All taxiways have a hold short line across them. They are designed similar to that on a roadway, and you never cross the solid line without stopping, and alongside the solid yellow line is a dashed line, which you taxi past. Since we cannot go onto the runway without stopping, the side toward the aircraft as we approached was solid. This is also the area where we do what is commonly referred to as the run-up. The run-up is the point that we follow a checklist designed for each airplane model and make, verifying that all systems are operating properly. Each cylinder has a left magneto and a right magneto, and we ground each side after running the engine up to about two thousand RPMs. When we ground the left, we will notice a reduction in RPMs and then when we turn both sides back on the RPM's will return to the two thousand RPM level. We then ground the other side and once again should see a drop in RPMs, which indicates that both sides are operating as they should, and we then go back to both to once again verify two thousand RPMs.

The next item on the checklist is to confirm the carburetor heat is working. It is a lever that moves up and down on the panel, and when in the up position carburetor heat is off so when in the down position it is on. As we move the lever downward, turning on carburetor heat, we should notice a reduction in RPMs. Most older general aviation aircraft have carburetors on them instead of fuel injection systems, similar to older road vehicles.

One of the key principles of a carburetor is the venturi, a constriction inside the carburetor that causes air to flow faster. We know that the faster a gas moves, the lower the pressure. This drops the air pressure in the venturi, which allows fuel to be sucked from the float bowl through the jet system. The drop in pressure also lowers the temperature of the intake air. This temperature drop is aided by the evaporation of gasoline, which is being atomized by the carburetor to burn in the engine. This temperature drop just past the venturi can be substantial. In airplanes, carburetor ice can form up to 90°F ambient temperatures, and a carburetor can theoretically drop the temperature by up to 100°F. This temperature drop cools the body of the carburetor. When the carburetor body drops below the freezing point (32°F or 0°C), any moisture in the air that hits the carburetor body has a chance of freezing onto it. Unfortunately, the temperature drop also tends to condense water vapor out of the air, just like water collecting on the side of a cold glass. If there's any water in the fuel, this can contribute to the ice problem. As this occurs, more and more water can freeze to the carburetor, which will gradually build up ice on the carburetor surfaces. The more ice there is, the less air can get through. This is the same principle that makes your refrigerator or air-conditioner work. So, in short, carburetor ice can be dangerous and can stall an engine.

In an airplane, heat is redirected to the carburetor, which warms the carburetor and melts the ice. Low-power settings and warm humid days are when carburetors are most likely to experience icing. When carburetor heat is applied, the first noticeable indication it is performing as designed is a slight drop in RPMs. As part of the preflight procedures of N1XF, I will look for this drop in RPMs to verify the carburetor heat is working.

As I move down the checklist, the next item is to verify good oil pressure by monitoring the oil pressure gauge; then the oil temperature, verifying the vacuum pump is working by checking the vacuum gage for proper pressure; and finally verifying the electrical system is charging. I will then bring the throttle back toward me to the closed, or off position, as a check to be sure the engine will idle without shutting down. This step is often overlooked but is extremely important since, on final approach to landing, the throttle is often in the closed position as the airplane will generally glide to a landing. Should I need to execute a go-around or climb-out for any reason, I want to be sure I will have the power available to safely execute the procedure.

On this day, everything checks out normally, and I determine the airplane is airworthy for our flight to Syracuse and back to Bradford. The final items on the preflight checklist is to turn the transponder to the altitude read position, which will enable air traffic control to read my altitude for separation from other aircraft. Since this will be a flight operated under (IFR), I will be assigned altitudes by air traffic control even though I have already filed a flight plan with them with requested altitudes. Normally, requested altitudes and headings are authorized in the clearance received from ATC, but they may determine it is not possible for safety reasons.

On a clear day with near unlimited visibility like this day, I have the option of departing under visual flight rules and then opening my instrument flight plan in the air or I can call the local center, in this case Cleveland Center, and request a clearance to depart within a small time frame, usually about ten minutes. If I open before departure, I will need to call Cleveland Center and wait a few minutes while they determine a squawk code and activate my flight plan.

I turned to Pat and said, "I think I will depart VFR and open in the air so we can save some time."

"That sounds good to me," he replied. "Would you like me to work the radios while you fly?"

"Sure you can do that," I said. "Chandi, we are ready to back taxi, are you still buckled in and ready?" I asked.

"Yes, I am, Dad," Chandi replied.

"Okay, Pat, we are ready to radio back taxi on Runway Three-Two," I said as I heard Pat start to say. "Bradford traffic, November One X-ray Foxtrot on taxiway Alpha, back taxing on Runway Three-Two for immediate departure to the Northeast, Bradford."

I gently pulled back on the yoke once again and slowly pushed forward on the throttle lever while releasing the foot pressure I was holding on both break pedals, sliding my feet to the lower half of the pedals, resting my heels on the floor for maximum control of the rudder pedals. N1XF slowly started to move forward toward the runway as Pat and I both scanned the pattern areas and approach course for any unknown aircraft in the air or on the runway surface. "All clear," I stated as Pat confirmed with me it was safe to enter the runway area. I applied pressure to the right rudder pedal as I slowly taxied the Comanche onto the runway surface, being careful to stay on the painted center lines.

At some airports, it is required that you have permission to taxi to another location while on the surface. These are referred to as controlled airports. Controlled airports are only those with active control towers on the field, which Bradford Regional Airport never has; therefore it is an uncontrolled field. It is the responsibility of each pilot in command to ensure the safety of everyone in an uncontrolled environment.

So as I am back taxing on the runway, both Pat and I are looking for traffic that may not be conscious of where we are and what our intentions might be. When back taxing on Runway 32 at Bradford from taxiway Alpha, there are not any other points of entry onto the runway from the surface, so our attention is on the final approach to the runway and to our right where the inbound traffic is likely to be approaching on the downwind. It is important to be alert at all times since radio communication is not required at uncontrolled airfields such as Bradford. In fact, although it isn't likely, it is possible an aircraft could be landing without any communication equipment on board since it is not required in the airspace around the Bradford Regional Airport. The airspace at Bradford is Class E airspace to the surface directly over the airport up to eighteen thousand feet Mean Sea Level (MSL). There are visibility requirements and distance from cloud requirements for VFR flights, but again, no radio or equipment requirements for VFR flights in Class E airspace.

As I approached the number "32" painted on the runway, I applied slight rudder pressure on the left rudder pedal, gradually moving N1XF to my left and to the edge of the runway so I could apply heavy right rudder and turn the airplane to the right by 180 degrees while making one last scan of the downwind and final legs of the traffic pattern for any incoming traffic. As N1XF turned to the right, it rolled into takeoff position with very little runway behind us and over six thousand feet of paved snow covered runway ahead of us. The nosewheel was lined up between the three and the two making up the number thirty two as it should be.

Runways are numbered relative to the magnetic direction they lay, rounded to the nearest ten degrees, and then dropping the zero at the far right of the number. Runway 32 at Bradford is magnetically 322 degrees. Rounding that to the nearest ten degrees would make it 320 degrees, so drop the "0" and you have 32.

Just prior to adding power, I adjusted my directional heading indicator (one of six primary instruments) to read three hundred twenty two degrees which I verified with the magnetic compass hanging from the front window frame on N1XF. Directional heading indicators operate with the vacuum system and need to be aligned with the compass approximately every thirty minutes. I then verified the trim wheel to be in the neutral position for takeoff, checked to be sure the carburetor heat was up and in the off position, pushed the red mixture lever to the right of the throttle all the way forward, which would put the fuel/air mixture in a full rich position, which is recommended for takeoff at Bradford, verified the electric fuel pump to be on. There are two fuel pumps on most single engine Piper aircraft, one is engine driven and the other is electric. If the engine-driven fuel pump fails on takeoff, I will have the electric pump on already, and it will instantly start to supply the needed fuel to the cylinders. I then glanced down and checked to be certain the fuel tank selector was pointed to one of the inboard tanks. All take offs should be with the fuel being supplied from the fullest tank and both inboard takes were at the full thirty-gallon capacity.

N1XF is now in position for takeoff and ready for what is to be an eventful flight and the final takeoff it will ever make.

THE FIRST 154 MINUTES OF FLIGHT

The time was now 11:33 EST, and we were ready to roll. With Pat working the radios, I instructed him to announce that we were starting our takeoff roll. Pat said, “Bradford traffic, November One X-Ray Foxtrot is rolling on Runway Three-Two for immediate departure to the northeast, climbing to seven thousand, Bradford.”

As Pat made this radio call, I reached over to the throttle lever and pushed the throttle full open as I again slid my heels to the floor and off the brake portion of the foot pedals. N1XF was rolling and gaining speed as I glanced at the gauges to verify all were operating in the green range, which would mean everything was operating normally. The manifold pressure would go to about twenty-five inches and the RPMs were around twenty-five hundred as we rolled down the runway, inching closer to being airborne. If there were any indication of malfunctions at this point, we could still abort the takeoff with over five thousand feet of runway still ahead of us.

"All in the green," I announced as Pat replied, "All in the green," confirming my interpretation of all the instruments reading normal; we were ready to start our climb from 2,143 feet to 7,000 feet MSL.

As our speed approached 85 miles per hour, I rotated gradually by applying back pressure (pulled back toward me) on the yoke, which changes the angle of attack on the elevator, the horizontal wing-type surface on the tail of the airplane. This will help to rotate the nose up more and is what is referred to as "rotation." As I rotated, I could feel the airplane getting lighter as first the nosewheel, then the main wheels under the wings left the ground, and we were airborne. As we climbed through one hundred feet above the surface, I reached over with my right hand and clutched the landing gear lever with the white wheel attached to the end of it, pulled out and up as I heard the gear start to retract, the main gear under the wing folding into the underside of the wings and the nose tire disappeared into the underbelly of the cowling area. I glanced out at the small mirror attached in a position so I could verify the gear was up and quickly scanned the area around the plane for any air traffic I wasn't aware of.

My target indicated airspeed for climbing was ninety-five miles per hour so I pitched the airplane by applying slight pressure on the yoke and saw the airspeed indicator move toward ninety-five. I reached down and started to slowly spin the trim wheel to adjust the trim tab on the trailing edge of the elevator. Without getting into aerodynamics and technical reasons for a trim tab, you only need to know that it will assist the pilot in maintaining the desired angle of attack as the aircraft glides through the air. In this instance, I was holding pressure on the yoke to maintain my desired indicated airspeed of ninety-five miles per hour; by adjusting the trim, I am able to release the manual pressure needed, and this adjustment would hold the yoke at my desired position. I then would only need to apply very slight pressures to the yoke to maintain my desired rate of climb.

I reached over to the power handle and pulled back, bringing the manifold pressure back to the recommended climb power of twenty-three inches then reached over to the lever with the blue handle, began to turn the round knob to the left, which would change the angle of the propeller blade to get us to my desired RPMs for climb of twenty-three hundred. We were now established in a constant controlled climb, climbing at an impressive rate of over one thousand feet per minute. Our Piper Comanche is an air-cooled engine, and the performance on this day was amazingly strong due to the extremely cold air.

As we climbed through three thousand feet MSL, about nine hundred feet above the terrain below us, I asked Pat to announce our intentions to make a right turn toward our heading of sixty-five degrees magnetic. I glanced at the instrument panel again to verify all gauges were indicating in the green before I left the area and must have taken longer than normal since I heard Pat say, "Are you all right?" which I replied, "Yes, I am, and everything looks great." Pat smiled and pressed the push to talk button on his yoke and began to say, "Bradford traffic, November One X-ray Foxtrot on departure from Runway Three-Two, turning right to a heading of zero-six-five, climbing through thirty-two hundred for seven thousand, Bradford."

As I slightly pressed on the right rudder pedal and turned the yoke to the right, N1XF began its turn to our desired heading, rolling to the right in a standard rate turn as it responded to my control inputs.

"Pat," I said, "let's contact Cleveland Center on one two four point three two [124.32] and get our IFR clearance." I glanced back at Chandi and noticed how she was intently peering out the side window, enjoying the beautiful winter view of the Allegany Forest below us.

Pat entered the 124.32 on the standby side of our KX155 communication radio and pushed the button to activate the frequency. "Cleveland, November One X-ray Foxtrot, five miles Northwest of Bravo Foxtrot Delta (BFD) with a request," Pat announced.

"Go ahead, November One X-Ray Foxtrot" Cleveland replied. "We would like to activate our IFR flight plan to Sierra Yankee Romeo [SYR] at this time," Pat said as we waited for Cleveland Center to respond.

"November One X-ray Foxtrot, radar contact two miles northeast of Bradford, squawk 4268 and ident" was the response from the controller at Cleveland Center. Pat replied, "Squawk 4268 and ident for November One X-ray Foxtrot." The controller from Cleveland Center responded by saying, "Read back is correct, One X-ray Foxtrot. Climb and maintain seven thousand, proceed on course." Again Pat replied to Cleveland Center by saying, "Climb maintain seven thousand, proceed on course, One X-ray Foxtrot." We now have our clearance to fly our flight as I filed it with air traffic control before I left my house a few hours ago.

This communication does seem to be elementary and redundant; nevertheless, it is a required step to be absolutely positive that the message is relayed and understood correctly. With the pilot in the aircraft reading back all instructions, the possibility of miscommunication is very small. If there is a directive that is not understood, the center and the pilot will almost certainly notice it immediately, along with having ample time to correct it for the safety of everyone using the system at the time.

For the first time since arriving at the airport Pat, Chandi, and I now have a couple minutes to relax and enjoy the flight. As we engage in small talk and general conversation, Pat and I both continually scan the sky for any other air traffic that might be around us. One of the huge advantages of flying under IFR rules is the ATC controller is sitting in the control room watching all flights in his control area on his radar screen. As other air traffic moves around, the controller will alert you to any air traffic nearby. Sometimes, it is other IFR traffic and other times it is air traffic flying under VFR rules. There are many times VFR traffic will show on the radar screen without altitude read equipment operating, and ATC will alert you to the traffic and state that the altitude is unknown. If the VFR traffic hasn't filed a flight plan with ATC, which isn't required, then the air traffic controller won't know where the VFR traffic is flying to, only it's current course of flight. All VFR traffic outside of strictly controlled airspace (Class B and Class C) will use a squawk code of 1200. So what a controller might see on his or her radar screen is all aircraft with assigned squawk codes, and many aircraft all squawking 1200. An assigned squawk code makes it much easier for a controller to know which aircraft is you and what your intentions are.

We are now about six minutes into our flight and on course toward Syracuse on a heading of zero-six-five as we approached our destination altitude of seven thousand feet MSL. About five hundred feet before we reached our assigned altitude, I began to push the yoke forward, gradually adding more and more pressure as the nose of N1XF began to lower and our climb rate started to drop. As we reached seven thousand feet, I reached for the power lever and pulled the power back to about twenty-one inches of manifold pressure and began to spin the trim wheel forward to trim off the forward pressure I am holding on the yoke. Once stabilized, the air speed climbed to about 165 miles per hour, which is within the desired range for cruise. Once again, I reached for the blue knob and began twisting it to the left as the RPMs started to drop back toward the twenty-one hundred number recommended for cruise flight. After verifying the RPMs to be constant at 2100 and the manifold pressure was steady at twenty-one inches, I began to lean the air fuel mixture by slowly pulling the mixture lever back toward me. The engine mixture should be leaned to just on the rich side of what is referred to as peak. As you start to pull the mixture back, you will see the RPM begin to rise a little. You will need to continue slowly pulling the control back until the RPM begins to drop. Then begin slowly pushing the mixture control forward toward the rich setting once more. Locate the point at which the RPM is the highest. This is the optimum mixture setting and just on the rich side of peak.

I verified the altimeter setting and reviewed the cruise checklist one last time before storing it in the cloth pocket by my left leg. The time is now eleven forty-seven and we are in great position about one mile above the terrain below us. With near unlimited visibility, it was as beautiful as I had envisioned the moment I had jumped out of bed.

"November One X-ray Foxtrot," I heard in my headset, "traffic eleven o'clock, eight miles at six thousand feet on a Southerly heading."

"You've got to love ATC," I said to Pat as he pushed in on his little red button to activate his push to talk switch.

"We have him in sight," Pat replied as he pointed in the direction of the twin-engine airplane approaching us from our left but well beneath us.

"That's what I brought you for, Pat" I joked as we watched the Piper Aztec cross under and by us.

At seven thousand feet and with near unlimited visibility, I could see we just went past Wellsville, New York, on my left; and Hornell, New York, was quickly coming into clear view. Pat and I chatted a little about our experiences with primary training, and I relayed to him what fun it was to fly in a Piper Tomahawk, a small single-engine, two-seat trainer manufactured by Piper. Flying past Wellsville had brought back several memories of some of my earliest flight lessons. I had experienced my first spin maneuver in the Tomahawk owned by Wellsville Flying Service.

It didn't take long until we were between Hornell and just south of Penn Yan and Kueka Lake was well in our sights. Keuka Lake is one of the smaller lakes in the Finger Lakes Region of New York State. The Finger Lakes are a pattern of lakes in the west-central section of upstate New York. The lakes are long and narrow (resembling fingers) and are oriented roughly on a north-south axis. The two longest, Cayuga Lake and Seneca Lake, are among the deepest in America. On this cold winter day, you could see the snow-capped hills with the lakes nestled in the valleys.

As we approached this area, I asked Pat to let ATC know we wanted to enter a holding pattern over the intersection "BEEPS." The air traffic system has what is referred to as airways, and they are systematically numbered. When two or more airways intersect forming an intersection, they are given five-letter identifications.

A holding pattern for instrument flight rules aircraft is usually a racetrack pattern based on a holding fix. The primary use of a holding pattern is delaying aircraft that have arrived at their destination but cannot land yet because of traffic congestion, poor weather, or runway unavailability. Several aircraft may fly the same holding pattern at the same time, separated vertically by one thousand feet or more. This is generally described as a stack or holding stack. As a rule, new arrivals will be added at the top. The aircraft at the bottom of the stack will be taken out and allowed to make an approach first, after which all aircraft in the stack move down one level, and so on. Air traffic control will control the whole process, in some cases using a dedicated controller for each individual pattern.

As an IFR pilot, I need to be proficient with flying holding patterns as well as all other maneuvers. Pat made his radio call, and ATC gave us our clearance to fly the holding pattern at BEEPS. I entered the hold over the fix on what is referred to as a direct entry and turned right to correctly fly the holding pattern, making the needed adjustments for wind correction to stay in the pattern, being careful not to let N1XF get blown off course. After two complete trips around the holding pattern, we were ready to depart and continue toward Syracuse Hancock International Airport.

We departed BEEPS intersection toward SYR on the Victor 84 (V84) airway as directed by ATC. We had been in the air approximately one hour now, and the flight was as smooth as I had anticipated and planned for. N1XF was performing remarkably well, and it was evident the atmospheric conditions were ideal for flight this day.

I chose Syracuse to practice at for a couple of reasons. The New York Air National Guard facilities are at Hancock Field. There is a certain air of force and mystic with military aircraft, and I love to see them and fly among them. And flying to Syracuse would allow me to fly toward what would be my destination the following Monday.

I reached over to the panel and turned on the electric fuel pump just before reaching toward the floor between my seat and Pat's seat.

"What are you doing?" Pat inquired as I put my hand on the fuel selector lever located on the floor between our seats.

"I am going to switch to the other inboard tank and draw fuel from there for an hour," I answered.

Pat glanced over toward me and asked, "What is the reason for switching tanks now? We have plenty of fuel yet where we are at?"

"I want to control the fuel in each tank so I know what I have left based on fuel burn, and it will keep our weight evenly distributed," I said. "It's the way I have been taught to manage fuel in this airplane and the system I am used to," I continued as I asked Pat. "You don't think I should?"

"Oh no, I am not saying that," Pat replied, "but do you really think that little bit of weight will make a difference in how this airplane flies or handles?"

"I suppose not, Pat," I said, "but I think it is a good habit to follow."

"It sure is, and I am not saying it's wrong, I was just wondering," Pat was saying as we heard the ATC controller say, "N1XF, what approaches would you like at Syracuse?"

It was time to prepare for our instrument approaches into Syracuse. We would not be stopping. We would be executing what is called touch and goes, which is when our wheels touch down on the runway and we then immediately add power and take off again. This way we could save time and still accomplish the goal of flying full approaches to the runway.

We had planned the two approaches I wanted to practice at Syracuse, and they are the localizer approach to Runway 28 and the ILS approach to Runway 28. "Pat, I would like to fly the ILS approach first and then get vectors back around for the localizer approach," I said.

"We would like to have vectors for the ILS approach to Runway Two-Eight," Pat informed the controller, "and then we would like to be vectored back around for the localizer approach for Runway Two-Eight." By this time, we weren't with Cleveland Center any longer, we were being controlled by New York Center so New York gave us our heading to turn to for vectors for the ILS approach for Runway Two-Eight in Syracuse and cleared us to four thousand MSL. I reached over and brought the manifold pressure back to seventeen inches and N1XF began a nice steady decent of about five hundred feet per minute down to four thousand MSL before stabilizing there as I brought the manifold pressure back into cruise setting.

An instrument landing system usually has two stations; one is the glide-slope and the other the localizer. The glide-slope is used to make sure the plane is on the proper decent rate toward the airport, and the localizer is to make sure the plane is properly aligned with the runway and heading the right direction toward the airport.

"November One X-ray Foxtrot contact Syracuse approach on one thirty-four point two seven," I heard in my ear and Pat read back, "Syracuse Approach on one thirty four point two seven for November One X-ray Foxtrot, have a good day." Pat then reached over to our communication radio and entered 134.27 and called Syracuse approach saying "Syracuse, November One X-ray Foxtrot with you level at four thousand with information Whiskey." The local ATIS (automatic terminal information service) is where we get ground conditions, winds, and other information about the airport from, and the most current conditions was called at this time whiskey. By saying we had information whiskey, approach would know we had the latest information available to us. "N1XF fly heading zero-seven-five, descend and maintain three thousand for vectors to ILS Runway Two-Eight" was the response from Syracuse approach.

"Zero-seven-five for One X-ray Foxtrot," Pat replied as I banked N1XF left toward a heading of seventy-five degrees and brought the manifold pressure back again until we reached three thousand MSL where we leveled off once again.

I turned to Pat and said, "The ILS frequency is one zero niner point niner and the inbound course is two eight zero," and Pat reached over and inserted 109.9 into our navigation radio and spun the reading to reference 280 degrees. ATC will move us around to the southeast of the airport, and once we are past the airport, they will have us make a left turn to intercept the inbound heading of 280 degrees. I handed Pat the approach plate and asked him to insert the frequencies for the tower and the local ATIS once again so I could listen to the surface conditions and wind directions one last time before approaching.

The Syracuse airport and Runway 28 was now a fair distance off our left wing and well behind us as the approach controller said, "November One X-ray Foxtrot turn left to heading three one zero and intercept the inbound course for Runway Two-Eight." Once again, Pat was on the radios as he read back our instructions, and I again banked left to a heading of 310 degrees, and N1XF was flying toward the localizer which would align us up to Runway 28 for our final approach.

On an ILS approach, the navigation indicators come from the VOR instrument in the airplane. There is a vertical needle, which tracks the localizer when the ILS frequency is active. The goal is to keep the vertical needle centered that is on course. If the needle moves to the left, the localizer is to the left, and you would be right of course; likewise, if the needle moves to the right, the localizer is to the right, and you are left of your approach course. There also is a horizontal needle, and the goal is to also keep it centered also. If the needle drops below the center mark than the proper glide slope to the runway is below you and you are high of the proper glide path. And if the needle moves above the center line, then you are below the proper glide and are too low. This is the primary navigation instrument to follow on an ILS or localizer approach until you reach an altitude that is designated on the approach chart as a minimum altitude, or until you have the airport environment in site, at which time you may fly visually and continue your descent to the runway. If the runway isn't in sight at the minimum altitude by what is called a missed approach point, you must start a climb and fly a published missed approach path until ATC provides further instructions.

As my vertical needle comes alive and starts to move toward the center line, I know that I am about to intercept the inbound course to Runway 28. I slowly bank N1XF into a standard rate turn and watch as my localizer needle centers while I level the wings, and we are right on course. I may need to fly a slightly different magnetic heading to correct for a slight cross wind so we don't get blown of course, but regardless, the needle needs to stay in the center so N1XF stays aligned with the runway.

N1XF is now established on the inbound course to Runway 28 as the controller with approach said, "November One X-ray Foxtrot, contact tower on one two zero point three."

Pat quickly replied, "Tower on one two zero point three for One X-ray Foxtrot." A push of the button and 120.3 was active on our radio, and Pat said, "Syracuse

Tower November One X-ray Foxtrot on the ILS approach to Runway Two-Eight."

"November One X-ray Foxtrot cleared to land Runway Two-Eight. I understand this will be a touch and go" was the response from tower.

"That is correct," Pat replied, "and we would like to be vectored back around for a localizer approach for Runway Two-Eight."

"Touch and go approved." The controller came back with "Upon departure fly runway heading, climb and maintain three thousand until further notice." Once again, Pat did the read back correctly, and we were cleared to land.

I quickly reviewed our prelanding checklist for the second time. The fuel selector was on the fullest tank, the electric fuel pump was on, and I reached over and grabbed the little wheel and pulled down on the landing gear lever. The landing gear started to drop, and as the green indicator light started to glow, indicating the landing gear had locked into place, I glanced out and saw the nose gear in the mirror, which verified the gear was in position. The mixture had already been moved to the full rich position as required, and I had turned on the landing light on my first check. I reached for the blue knob and pushed it forward, which changed the angle of the propeller blades for maximum RPMs.

One last item before touchdown was to verify everyone was still properly buckled in. I could see Pat was, so I said, "Chandi, are you still buckled in?" She gave a quick yes response as I pulled up on the lever, lowering the final portion of flaps. I started to pull the throttle back to idle slowly as we gently glided over the threshold. The left wing was slightly lower than the right to compensate for the wind as N1XF gently settled onto the left main wheel then the right before the nosewheel touched down.

A quick adjustment to the trim wheel, a gentle pull and release of the flap lever to raise the flaps, a quick glance over to the carburetor heat to verify it was in the off position, and I was ready to push the power throttle forward to once again apply full power to our 260-horsepower engine. N1XF quickly responded and we were on the roll again ready to fly runway heading and climb back to three thousand feet MSL.

Our second approach would go much like the first, and the vectors were very similar as were the radio calls. As we climbed out from our second approach, we were cleared to the requested altitude of six thousand feet MSL for our return back to Bradford. Our on-course heading would be about 245 degrees. We quickly found our course and started our climb to our cruise altitude of six thousand feet mean sea level when N1XF "hiccupped," an ever so slight miss that was defined enough for Pat and me to glance at each other, and Pat asked, "Did you hear and feel that?"

"Yes, I did, Pat," I replied as Pat quickly started a scan of all gauges, looking for any sign of problems. "Everything looks great, and it seems fine now," Pat stated as he continued to monitor the performance of N1XF.

"I agree, Pat" was my response. "Let's continue and we'll keep a close eye on it and divert to another field if we need to." I continued, "I will tell Bruce about it when we get back on the ground in Bradford and see what he thinks."

On our trip to Syracuse, our assigned altitude was seven thousand feet, but our return was at six thousand feet. IFR flights are always flown at one thousand foot intervals with flights on an easterly heading assigned the odd thousands, and those on a westerly heading are all assigned the even thousands. Our choices on our return trip were four thousand, six thousand, or eight thousand, and we had requested six thousand. By flying at six thousand, we would be high enough to have some altitude in case of an emergency, which would give us additional time to trouble shoot or to locate a suitable landing spot (in comparison to choosing four thousand feet). If we flew at eight thousand feet, we would experience a stronger head wind on our trip back to Bradford, which would slow our groundspeed considerably.

Once at altitude and set up for cruise flight on a clear and smooth flying day like this, the largest responsibility a pilot in command has is to monitor the aircraft, maintain heading and altitude. This time is usually used for small talk, planning the approach stage of flight, monitoring other traffic when possible and other tasks that may occur during flight.

As instructors, Pat and I are generally in the training mold and this day wasn't any different. As we were flying across the southern tip of the Finger Lakes Region and admiring the scenery far below us, Pat asked, "If you lost your engine right now, where would you go?" As an instructor, it is a common question to ask, and as a pilot, I should always be aware of the terrain below me for this purpose. I quickly responded, "Right there off my left wing tip is a nice long field that seems to be wide open without any obstructions. I think I could make it there easily if I needed to."

"Good choice." Pat chuckled. "That is the field I had in mind."

It seemed like only a few minutes, and we were flying past the Wellsville airport. This time, it was off our right wingtip as we were on a southwest heading and maintaining our altitude. It had been a few minutes since New York Center had handed us off to Cleveland Center, and we were now only several minutes from starting our descent to the Bradford Regional Airport. In a matter of a few minutes, we were flying over Port Allegany, Pennsylvania, and off our right wingtip, you could see the area that I often take my students to when we were practicing maneuvers. I pointed the area out to Pat as our practice area and showed him some of the landmarks that make it an easy area for students to indentify from the air. There is a noticeable stream running through the farmland, one that winds into the unusual shape of a horseshoe. In this same area is an asphalt and gravel plant all tucked between the hills. The south end of the valley is the town of Port Allegany and to the north is the small Borough of Eldred.

"November One X-Ray Foxtrot, descend and maintain four thousand, proceed direct to Bradford" was the radio call from Cleveland.

"November One X-ray Foxtrot to four thousand," Pat quickly responded as I now could see the Bradford airport off in the distance. We were about eight to nine miles northeast of the field and started to plan to fly the VOR approach to Runway One-Four. "Cleveland," Pat said, "we will fly the VOR approach to Runway 14" as I reached for the throttle handle to bring the power back to seventeen inches of manifold pressure, which will start our controlled descent. As I let go of the throttle, I verified seventeen inches of power and watched as the needle stabilized where it should.

As I pulled my hand from the throttle and back to the yoke, I was looking around the area for other traffic I could hear on the radio. The time was 2:07 EST; the moment something began to go terribly wrong and this beautiful winter day became our great adventure.

THE FINAL EIGHTEEN MINUTES

The first indication of trouble began as the engine in N1XF began to hiccup a little, almost like the cylinders weren't firing properly. I immediately glanced down to the manifold pressure gauge as my hand shot back toward the black throttle knob, verifying that I had not accidently brought the red handled mixture lever back toward me, which would lean the fuel and air mixture to the point of becoming to lean to burn.

The manifold pressure was rapidly dropping along with the RPMs as the engine began to struggle more and more. Within seconds, it was obvious that we were in an emergency and were losing total power. The propeller continued to swirl in front of N1XF, which should have given us the spark we needed to restart the engine.

"Pat," I anxiously and emphatically said, "we are losing total power. It sounds like the engine is starving for fuel, but we have plenty of fuel on board. The fuel gauge is indicating about half full."

"Get to best glide, get to best glide right now," Pat sternly said. "What is best glide speed?" Pat demanded.

"I am almost there, Pat," I responded. "Best glide is 105 miles per hour. We are there now, holding 105." I continued by saying, "Pat, I have the emergency checklist. You trim this thing to maintain 105 and fly it. I am on the emergency list."

"I have the plane," Pat said. "I have the plane."

"You have it, Pat" was my response to Pat. "Make the radio calls, Pat," I continued. "Radio an emergency and let ATC know we are in an emergency here."

"I am on it" was Pat's response as he pushed his finger on the push to talk button saying, "Cleveland, November One X-ray Foxtrot declaring an emergency." Pat hollered into his headset. "We have lost total power and are declaring an emergency."

Pat reached over to the transponder and turned each of the four small knobs so we could squawk 7700. By squawking 7700, we would be easy to pick out on the controllers screen and the code 7700 is the emergency squawk code.

"November One X-ray Foxtrot, I understand you are declaring an emergency at this time" was the confirmation from Cleveland Center.

"That is right," Pat shot back. "November One X-ray Foxtrot, the area is clear to fly as needed. If I can assist, in any way let me know," the controller anxiously responded.

"What is going on, what happened?" Chandi quizzed from her back seat.

"We have lost complete power, Chandi," Pat replied. "We are trying to get restarted."

"I know you are busy," she said. "I will keep quiet."

"Just make sure your seatbelt is buckled and tight," Pat instructed her as he looked around out the windows.

"I am looking around for a field, Steve, I don't see any emergency landing options here," Pat said.

"Doesn't look good, does it?" I responded. "It's all trees for miles here. We need to try making the airport."

"Yes," he replied. "I think we need to try a right base for Three-Two," Pat said.

"Okay," I answered as Pat inserted 123.07 into the communication radio and activated it. The Bradford Unicom frequency is 123.07, and as soon as Pat had activated it, I heard him saying, "Bradford traffic, Bradford traffic, November One X-ray Foxtrot is declaring an emergency. We are about six miles east descending from four thousand. November One X-ray Foxtrot is on an extended right base for Three-Two. I repeat, November One X-ray Foxtrot with an engine out declaring an emergency landing, approaching on a right base for Runway Three-Two."

Chandi sat quietly in the back seat with an alarmed and concerned look on her face as I systematically worked through the emergency engine out checklist.

First by applying the carburetor heat just in case we had ice in the carburetor, which was unlikely, then by verifying the electric fuel pump was on, the mixture was in the full rich position, and the ignition was still in the on position. I then verified fuel was indicated in the active tank as I pumped the throttle, trying to get the engine to fire again. I reached for the fuel selector between my seat and Pat's seat and rotated it to the other main inboard tank and again attempted to restart the engine. I did this for each of the six tanks we had available to us to no avail.

As I rotated through each tank each time, I was verifying the magnetos were on as I tried to restart the engine. There is an emergency checklist that all pilots should memorize and be able to work with from memory. It is aviate, navigate, investigate, communicate, and secure. Before we moved to the secure part, I wanted to confirm with Pat that I wasn't panicking and was interpreting and recognizing all controls correctly. So I said, "Pat, I am going to go through the checklist once more with you. I need you to confirm everything and my interpretations as I move through the list."

"Let's do it," Pat said.

And so I started through each item, verbally announcing what I was doing and verbalizing what I saw. After each check, Pat would say "check" until we reached the restart attempt and Pat reached over and attempted a restart himself.

As I looked out in front of us, I saw the terrain getting closer at a rate that greatly concerned me. "Pat," I said as I, again, looked around for an alternative landing spot, "we are not going to make Three-Two. I think we need to attempt a straight in approach to Runway Two-Three." Pat was quick to agree with me and instantly was on the radio announcing, "Bradford traffic November One X-ray Foxtrot is straight in for Runway Two-Three. I repeat, we are on an emergency approach straight in Runway Two-Three Bradford."

Almost immediately, I recognized Brian's voice responding over the Bradford Unicom saying, "November One X-ray Foxtrot Runway Two-Three is closed and snow-covered. It has not been plowed."

"We have no choice" was Pat's quick response. "November One X-ray Foxtrot has an engine out emergency and will attempt to land on Runway Two-Three. We are coming in on Two-Three."

"Are you in need of emergency personnel?" Brian came back with adding, "Do you need the fire truck to respond?"

"We will need all emergency personnel and equipment available," Pat responded. "If we don't make the runway, we will be in the trees on final approach to Runway Two-Three. We have three souls on board," Pat added.

By this time, I realized we were either going to land in almost two feet of snow on Runway 23 or fly into the snow-covered trees approaching Runway 23 at the Bradford Regional Airport. We had unsuccessfully used every exercise available to us to restart the engine in N1XF. I made one last scan of the emergency checklist and finally relented and followed the final steps of securing the aircraft for a crash landing. I reached over and put the fuel selector to the off position, pulled the mixture knob back toward me into the lean position, and pulled the black knob on the throttle to the closed position. The purpose of doing this is to clear the fuel lines of any fuel and attempt to clear any fuel from the engine to lessen the possibility of an explosion or engine fire on impact. The last step was to turn the ignition to the off position, which would effectively shut off both magnetos to kill any spark. The final decision Pat and I made regarding the configuration of N1XF was to leave the landing gear in an up position. By doing so, it would create less drag, and if we flew into the trees, as was now obviously going to be the case, there would be less for a tree to catch on to turn N1XF upside down or spin us out of control. The only item on the checklist we missed was unlatching the door. By doing so, if the latch got bent and jammed, we would still be able to open the door to quickly escape.

For a few seconds, it was deafly silent in the airplane. We were resigned to our fate, and Chandi had been an incredibly brave lady as she sat silently in the back seat, listening and watching as Pat and I concentrated on our responsibilities. I slowly put the checklist in my left hand and moved it into the red leather and cloth pocket by my left leg. As I had glanced to my left to look down, I was able to see the treetops just off our wingtip. I distinctly remember calmly thinking and silently saying to myself, "Lord, these trees aren't all that big around, they are only tall. Maybe, just maybe we have a chance here."

As Chandi noticed we were now just a few feet above the trees, she asked, "Are we at the airport?"

"No," Pat responded. "We aren't going to make the airport. Buckle your seatbelt as tight as you can and put your head down." Chandi's only response as she calmly followed Pat's instruction was "Oh shit!" She later shared with me that her quiet time in the back seat was spent preparing herself to die and talking to God.

There is a saying in aviation among pilots. It goes something like this: If flying at night with your landing light on and you don't like what you see, turn the light off. If flying during daylight hours and you don't like what you see, put your head down and close your eyes.

We didn't like what we were seeing, but I couldn't close my eyes or put my head down. I suppose it's part of my competitive nature; I just wasn't going to give in or give up, even though I now knew we were going to crash into the trees at one hundred five miles per hour.

"Bradford traffic, Bradford Unicom," I heard Pat say as he broke the silence. "November One X-ray Foxtrot is going down, I repeat, November One X-ray Foxtrot is going down! We are going into the trees on final approach to Runway Two-Three." By this time, the runway was painfully close, less than a half nautical mile straight ahead of us. If we only had another one hundred feet of altitude, we could make the runway, I thought.

We could see the huge green Bradford Regional Airport fire and rescue truck rushing onto the runway, plowing through the snow as the snow sprayed high in the air to the side. Following closely behind were a couple of the other airport trucks. The fire and rescue truck is loaded with water and has a huge cannonlike fixture mounted to the top of the cab, capable of spraying a huge amount of water an impressively long distance.

I could now feel the tops of the trees start to scrap the underbelly of N1XF, and I hoped one wouldn't shoot through the thin metal between us and the treetops. It was too late to maneuver the plane so Pat and I both moved our hands off the control yoke, onto our laps as our feet moved to a flat position on the floor. Our shoulder straps were snug, the seatbelts tight around our waists, as we continued our fall into the trees. Pat did a marvelous job aviating while I was troubleshooting and executing the procedures on the emergency checklist. As N1XF fell into the trees, the fuselage was mostly aligned over a winding logging road, too narrow for us to land on and definitely not straight enough. To our slight right was a small body of water, I will call it a small pond. If we fell into the pond or flew into it, we probably would drown. As N1XF dropped below the treetops, each wing impacted some trees, one main tree on each wing.

The tree that struck the right wing was thicker and bigger than the tree that struck the left wing, and I believe the tree on the left wing broke off. N1XF started to veer to the right as the nose dropped toward the ground. As N1XF started to spin to the right, the propeller slammed to the ground and instantly stopped spinning. As the nose of N1XF hit the cold, frozen surface, the motor mounts broke and bent left as the entire nose section broke to the left. This is the one thing that saved our lives since in many crashes of this type, the nose section is pushed through the firewall into the cabin area, killing the occupants instantly.

As N1XF impacted the trees and started its spin to the right, the windshield started breaking into real tiny pieces and filled the air all around us. It's a vision I won't forget as those tiny pieces flew past my face and the shoulder strap strongly held my upper body back against the seat. It happened so fast, it was a mere few seconds, and N1XF came to a stop. We had gone from moving 105 miles per hour through the air to a sudden stop in just a couple seconds. The tail section was still in the air and the cabin area was pointing toward the ground, being held up by a few small trees and the thickness of the nose section that was pointing left, still partially attached to the airplane's firewall, and hugging the ground.

Everything had stopped by now. It was eerily quiet as I hung there in my seat for just a couple of seconds thinking about what had just happened. I knew I was alive but wasn't sure about Pat and Chandi. The accident had happened, and I hadn't felt any fear, but now, fear set in. I wasn't sure if I should look around toward Pat and Chandi, or if I should speak. I chose to speak!

WE'RE ALIVE. THANK GOD WE'RE ALIVE!

Looking straight ahead, I somehow garnered the courage to ask, "Pat, are you all right?" "Yes I am" Pat quietly responded. I took a deep breath and hesitantly asked, "Chandi are you all right?" What a relief when I heard my daughter say "Yes Dad, I am fine." "Thank you, God" I muttered and then turned to Pat and said "We need to get out of here fast before it catches on fire or explodes."

Pat reached his long arm over to the door handle and pulled the lever up, but the latch would not release the door to swing open. "The door won't open," Pat stated.

"Then we need to kick the window out and we need to hurry," I responded as Pat once again wrapped his strong fingers around the lever and pulled while the latch sprung loose and the door fell open.

Pat and I both reached for our seatbelt latch and unbuckled our straps and started to climb out the door. As Pat slid on down the leading edge of the wing to the ground, I reached for my gloves and worked my fingers into them, anticipating a cold long walk back to the airport. Pat is a tall man, so he stood in front of the wing to assist Chandi and me as we crawled from the wreckage. I climbed over the seat, out the door, and onto the wing as Pat assisted me to the ground.

"Come on, Chandi," I said, pressuring her to hurry. "I can't get out, Dad," she said, which startled me. "I cannot get my seatbelt unbuckled," she continued as Pat reached up into the wreckage and helped her. Pat reached into the airplane and helped her with one hand, physically lifting her as he helped her get out of the airplane.

Now we were all three safely on the ground, standing in front of the right wing as we turned toward the airplane to look at the mangled wings and the crumbled cowling. It seemed like an eternity but was only several seconds, and I said, "Let's move away from the plane before it catches on fire or explodes. We need to get as far away as we can." In agreement, Chandi, Pat, and I started to trudge through the knee-deep snow toward the logging road about twenty feet ahead of us.

Pat is a strong and tall man, so he moved faster than Chandi and me and at the top of his voice started to exclaim, “We’re alive! Thank God we are all alive.” Throwing his hands toward the heavens, Pat repeated himself a couple of times before reaching the logging road. He turned toward us as Chandi had her cell phone in her hand, trying to reach her friend, Burk. I reached over and put my arm around her back, and as I was looking into her beautiful blue eyes, I told her how much I love her.

She smiled and said, “I am sorry, Dad.”

“For what?” I shot back.

“For saying ‘oh shit,’” she seriously said. I started to chuckle at this and said, “You know what? It doesn’t matter what you said in that situation, you could have said anything. If God, or me, or anyone else couldn’t understand the situation, well then, I wouldn’t think there would be a God. But understand that God just gave us a huge gift.”

By now, the three of us were standing on the logging road, looking around, trying to figure how to find our way out of there and to the airport. We could see the opening leading toward the threshold of Runway 23 but between us and the runway was a tall chain link fence. The fence would be extremely cold, the snow was deep, and it would be hard for us to climb over it.

As we stood there trying to decide on a direction to walk, we heard the rumble of an excavator or other heavy equipment. Somewhat surprised that anyone would be logging or digging on such a cold day as today, we were excited to know that help might be close. Without a word, all three of us started to walk down the logging road toward the noise we were hearing. Several hundred feet further into the woods we were walking, and as we rounded the small bend, we could see a clearing where freshly dug dirt was piled. Just beyond this area was a large backhoe or excavator digging at the earth. "Look at that," I exclaimed to the others in amazement. "There is a truck and someone is operating that backhoe."

Trying not to startle the operator, we approached the heavy piece of equipment from the back and right, keeping our distance to stay safe. As the operator swung the bucket toward us, he looked up and noticed us standing there, waving frantically trying to get his attention. When he noticed us, he dropped the bucket to the ground and shut the machine off. As he jumped down off the seat high off the ground, we approached him. "We need your help," I stated while telling him that we had just crashed in an airplane just around the bend. "Could you give us a ride to the airport?" Pat asked while the operator stood there in amazement.

"Yes, I will," the operator responded."Are there just three of you or is there someone else that needs help?"

"Just the three of us," said Pat as the operator inquired to make sure we were all right.

"Let me clean off my seat and then get in," the equipment operator said as we all moved toward his red pickup truck parked by the entrance to his site. "It certainly is warmer in the truck, and I will take you to the airport," he offered as he began to ask us questions about what happened and where did we go down at. As the four of us crowded onto the bench seat in his pickup truck, we started to tell him how we ended up in the trees. He started his truck, and I immediately knew he hadn't been working there long as the heat blowing out of the heater vents on the dashboard of his truck was immediately warm and comfortable.

As we rode over the bumpy and ice-covered logging road, we rounded the bend we had just walked by and N1XF came into view on our left. "Oh my God," the gentleman exclaimed, "I cannot believe all three of you walked away relatively unharmed." We all were fortunate, and each of us had a very small cut on or close to our hands from the tiny flying pieces of the windshield. The cuts were very small, and only a very little bit of blood was visible on each of us. As we passed N1XF laying there partially on the ground and partially in the trees, the pickup driver just kept unbelievingly glancing over at the plane and then back toward each of us.

By this time, it had been several minutes since we had crashed, and as we were traveling over the logging road, approaching us was a yellow truck from the airport. Brian was driving it and rushing into the woods trying to locate us and the wreckage. As we approach the airport truck, both drivers rolled their windows down and Brian was somewhat surprised to see the three of us in the truck.

"Are all of you okay?" he asked as he reached for the handheld communication mic in his truck. "EMTs are on their way. You guys are real lucky," he told us as he keyed his mic and announced that we were all safe and on our way back to the terminal. The local ambulance and fire crews were told to meet us at the terminal as

Brian pulled away and quickly turned his truck to lead us back to the airport terminal building.

As we reached the secondary blacktopped road that would lead us to Route 59 and then back to the airport, an ambulance and a fire truck went flying by with the sirens blaring and lights flashing. We quickly fell in line behind them and the truck Brian was driving as we turned right onto the road. We were now out of the forest and safely on our way back to the airport. I reached down and pulled my gloves off my hands and thought I slid them into my coat pocket.

We were now on Route 59, leading us back to the service road onto the grounds of the Bradford Regional Airport. Soon we were turning right on to the service road and down the short road and around the bend, back by the parking lot I had met Pat at only a few hours ago. As the terminal building came into sight, we could see several fire trucks and rescue vehicles in the circle in front of the main entrance into the terminal. There must have been around twenty emergency personnel waiting for us to arrive.

As we pulled up to the circle and rolled to a stop, Pat opened the door; and Chandi, Pat, and I exited the red pickup truck, thanking the driver for helping us get safely out of the forest. Medical personnel immediately approached us; a few of them escorted Chandi into one of the ambulances parked around the circle. It is the last time I saw her for a few hours.

Pat and I were directed to the fuel office on the northwest side of the terminal building. We were surrounded by volunteer emergency personnel as we walked the short distance on the sidewalk, turned right along the terminal building to the steel door entering the fuel office area. Once inside, Pat insisted he was fine and walked over to the phone sitting on the counter saying, "We need to call NTSB (National Transportation Safety Board) immediately to comply with regulations and report this incident." I was directed to a chair behind the counter and sat down. I don't remember who I was talking to, only that I was being asked many questions about how I felt, did I hurt anywhere, had I bumped my head, was I aware of where I was, do I know what happened and several other medical questions. I remember an EMT taking my vitals like pulse, blood pressure, and more.

By this time, Pat was on the phone communicating with NTSB and answering their questions. I was trying to concentrate both on what Pat was saying and on what I was being asked. People were steadily coming in and out of the fuel office, and I heard the familiar voice of my cousin Ed. I had been the CFI (certified flight instructor) who had finished training Ed for his private pilot's certificate and endorsed him for his check ride. He had his certificate for a few months now, so I knew he knew me well, and knew how I thought as a pilot, how I flew, and how I follow procedures. Ed and I also are the same age and our families had been close while we were younger so we knew each other our entire lives. I felt a relief when Ed walked in, and I immediately assured him I was physically fine. Ed was also an EMT, and he calmly asked me a few questions. He had been at his house; about twenty miles from the airport when he heard the emergency call while listening to 123.07, the local traffic frequency for Bradford, and immediately knew I was the pilot in command of N1XF. He told me how he had dropped what he was doing and rushed to assist and asked if he could do anything for me.

My thoughts instantly went to my family. "Yes, I need you to call Mom and Dad," I said. "I don't want them to hear it from anyone but you or me. If they hear of an accident, they will know it is us, and they know Chandi is with me."

"I think Al is at the restaurant but I haven't seen him yet," I continued. "We've got to make sure he knows we're okay." Al is my brother who owned and operated the Runway Café, the restaurant located in the terminal building. "Just make sure you tell my parents you are here and you are with me. Assure them everyone is safe," I directed.

Pulling my cell phone from my left front pocket of my pants, I continued, "I need to call Joni, but I don't get cell service here." I flipped open my phone and noticed I had a couple of bars, so I dialed our home phone. As the phone rang a few times, I started to pray that my wife would pick up the phone. I felt she needed to hear my voice, not someone else, assuring her I was safe. After a few rings, the answering machine picked up, and I needed to decide if I should leave a message or try again later. She didn't have a cell phone so calling her cell phone wasn't an option. I decided to leave a short message since we live in a small area, and even though we lived in New York State and Bradford was in Pennsylvania, it was just over the hill and we know many people in both communities. In a small rural area like ours, word spreads quickly, and stories are often twisted with little detail. Often what is fact is far from the rumors that quickly start to circulate around the communities. I didn't want my wife to hear about the accident from anyone but me, and I certainly was concerned what she might hear would not be the facts.

As the EMTs were taking my blood pressure and other vital signs, I decided to leave a message. Not sure exactly what to say, I said, "I am at the airport and there has been an accident. We lost our engine and crashed into the trees, but everyone is fine. The EMTs are here and checking my vital signs, and I promise you, I am not hurt at all." I paused for a moment as I looked around the airport fuel office and listened to the conversation. Pat was still on the line with NTSB, Ed was standing there beside me, and there must have been ten to twelve emergency personnel just mingling around, not sure what to do. For a moment, it was almost like time was standing still. I mumbled into the phone to my wife, "Call Al at the restaurant, and he will talk to you."

As I put my phone on the counter, I couldn't be positive my call could actually be recorded due to the horrible cell service available at the airport. I could only hope that somehow Joni would hear my voice and know that it was okay. I was once again being peppered with questions like "What is your name?" "Do you know what happened?" "Who was with you in the airplane?" "Do you have any pain anywhere?" "How many fingers do you see?"

I looked out the fuel office window facing the tarmac and saw my brother quickly approaching coming from the restaurant. He opened the door and walked straight over to me, ignoring all those around him as they were telling him I was okay. After assuring him I was fine and that Mom and Dad had been notified, I asked him to call Joni again on the restaurant landline to try and reach her. I wanted him to tell her he had talked to me and to assure her that we all were fine. Just as quickly as he arrived in the fuel office, he departed to do as I had asked. Al later told me as he returned to the restaurant the phone was ringing and Joni was on the line calmly asking what had happened.

We had now been in the fuel office for a period of time as I looked around and suddenly realized that my daughter wasn't in the room. For the first time, I felt some panic and began to quiz the EMTs about the whereabouts of Chandi. I was confident she was physically all right as I had been with her for several minutes after the crash and remembered our conversations as we trudged through the snow together. But I began to become concerned that she was being held away from me for some other reason. I turned to the EMT standing beside me and asked, "Where is my daughter?"

He replied, "She is in the ambulance being attended to and is fine. We just need to be cautious and make sure she isn't injured."

I turned to Ed and said, "Ed, please go out to the ambulance and check on her for me. Tell her I will be out in a few minutes and Pat and I are fine. Let her know we have contacted her grandparents and mother."

It had been some time, and I could hear Pat concluding his conversation, and he soon hung up the phone and walked toward me. He stood beside me as he summarized his conversation with the NTSB and assured the emergency personnel that he was fine. The room was starting to thin out some as Pat began to wonder if someone could take him back to the crash site to recover his logbook. A logbook is extremely important to a pilot since it is the only place an accurate recording is logged of a pilot's experience. Pat had a sizeable logbook with many entries that were important to his flying career.

Brian offered to drive Pat back to the wreckage to see if we could locate Pat's logbook. I was growing impatient with all the commotion and attention and was ready to get out of the fuel office. Following Pat and Brian out of the side door of the fuel office to the truck, I quickly decided to find Chandi and bring her with us. As I rounded the corner, I could see her standing by the ambulance talking to EMT personnel and quickly got her attention and told her what where we were going and what we were going after. She had also been questioned by medical personnel and was determined to go with us. We crowded into the yellow pickup truck as Brian started it. We were soon retracing the roads that delivered us back to the terminal building on our way back to the crash site.

As we turned onto the muddy dirt logging road and approached the wreckage of N1XF, we saw a

Pennsylvania State Police car parked just a few feet from the crash site. The trooper was sitting in his car, and as we approached the site, he opened the driver's side window of his car and inquired why we had returned to the site. Pat explained that he wanted his logbook, and it had been in the back seat of the airplane as we crashed. The trooper's response was "This is a secure site. No one is allowed access to it until the NTSB or FAA arrives and gives permission."

This was my first indication of what was ahead of us. As the pilot in command of the flight, I was going to come under great scrutiny. Pat's decisions, training, and experience would also be under the microscope so to speak, and we needed to be extremely careful who we talked to and what we said. Even though we hadn't violated any federal aviation regulations and had relied on our expert training as we made decisions, it was immediately apparent to me that Pat and I needed to seek out legal representation. Thank goodness attorneys with extensive aviation experience were available to us at no additional cost through our membership in the Aircraft Owners and Pilots Association (AOPA) since we had purchased the legal protection plan through this reputable organization.

It was simply smart to remain quiet and wait for the advice of our legal representatives. And so we did!

TIME TO REFLECT AT THE AIRPORT

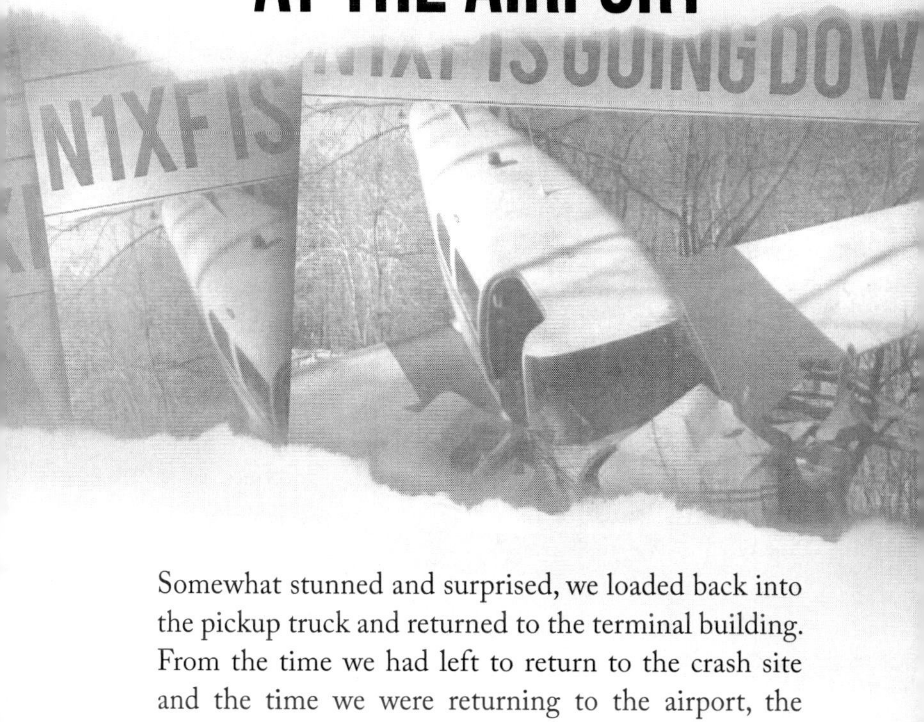

Somewhat stunned and surprised, we loaded back into the pickup truck and returned to the terminal building. From the time we had left to return to the crash site and the time we were returning to the airport, the emergency vehicles and personnel had left the area.

It was now mid- to late afternoon; Pat, Chandi, and I had some time to sit and talk about our day. Al had informed me that Joni was on her way to the airport and would be arriving soon. Pat needed to contact his family yet and inform them what had happened. It was unlikely they would be aware since Buffalo television stations had called the restaurant to inquire about the accident and seemed uninterested when Al had told them there were no injuries or fatalities.

As we climbed back out of the pickup and walked through the main terminal doors, there were few people around. The lobby is fairly small but well decorated with several chairs for passengers and others waiting for arrivals. Bradford Regional Airport is serviced by a commuter flight connecting with Pittsburgh. The few people there were waiting for arrivals and had heard about the crash there that day. As Pat turned right and approached the pay phone on the wall to contact his family, Chandi and I walked toward the restaurant doors. There was a husband and wife standing there in the lobby as we walked through. The older gentleman turned toward us and asked if we had heard there was a plane crash there a few hours prior. "Yes," I responded, "I was piloting the plane and my daughter here was with me. The three of us walked away unharmed, and that is all we can say at this time. Except," I continued, "God was good to us today."

Mid-January isn't a great time of year for a restaurant in Pennsylvania, and especially on a bitter cold Saturday evening on top of a hill, at a small rural airport ten to fifteen miles from the closest town of any size. As Chandi and I approached the double doors entering the Runway Café, I noticed how few customers were there eating. Al walked over to us and invited us to sit at any table. We looked around and chose to sit at a table by the large window, allowing for a good view of the ramp area leading to the runways. It was an area that was somewhat removed from the few customers eating at the time.

As Chandi and I sit down, we simultaneously glanced out toward the approach end of Runway 23. If must have only been a few short seconds, but it seemed like a long period of time that we both glazed out over the snow-covered runway. I am sure Chandi was replaying the events of the day in her mind as I was. Our thoughts were interrupted as we heard Al offer to get us a glass of soda and something to eat. I wasn't hungry and neither was Chandi, so we both replied, "No, thank you."

"Wow, Dad," Chandi said, "I cannot believe what we have been through together today. It seems like a dream, and we are now awake. I wanted to ask so many questions," she continued, "but I knew you and Pat were busy so I just kept my mouth shut and prayed. I didn't understand what you guys were doing, but I knew by the sound of your voices that we were in trouble. But the way you and Pat stayed calm was reassuring to me and helped me not to panic."

"We have practiced emergencies over and over in our training," I said. "Today, our training just took over. We didn't have to think about it at all, it just happened. By the time I was through the emergency checklist, we were clipping the trees. I am positive staying calm happened because I was busy trouble shooting." I continued, "And I never really thought about dying, it honestly didn't cross my mind at all. God was in control. I didn't think about praying either. I had this strange peace the entire time that we would survive. I just don't know how to explain it."

"I am going to go check on Pat," I said as I stood up. "Are you all right here, or do you want to come with me?"

"I am okay, Dad," Chandi replied. "I want to stay here and have a quiet moment by myself." I turned and walked toward the doors and back out of the restaurant into the lobby area. As I looked around, I noticed Pat talking on the pay phone across from the restrooms. It was apparent that he was in the middle of a phone conversation with his wife. He was quietly talking to her with his head down and his free hand cradling his forehead. Moments like this are to be private, I thought as I turned away to allow Pat the privacy he deserved.

I suddenly began to pace the lobby floor as my thoughts turned to my wife, Joni. I knew she was in route to the airport and wasn't sure what she had be told or heard. Times like this is when so much goes through our minds, memories made together, dreams shared with each other, good times experienced, as well as bad times endured. We all have them and have experienced them with someone. I didn't know what to expect from her when she arrived at the airport. She is both strong and weak. Many times I had seen her fret and worry over small, trivial matters while being a rock when I was sure she would be weak. In a lighter moment, she once asked me what I wanted her to tell people if I died in a plane crash. "Tell them I died trying to get high, legally," I often joked. Today I was in a plane crash but didn't die. I began to feel like I had stared death down and won, and I wondered if I would need to comfort her, or would she have to be my pillar of strength when she finally arrived.

I heard the sliding glass doors start to open, so I turned to look who was walking through them when I saw Joni calmly entering the lobby. I didn't notice any sign of tears, and she was amazingly composed. As we walked toward each other, I did notice her eyes had small tears forming. With a concerned and startled look on her face, she said, "Steve, I didn't know Chandi was with you. Where is Chandi?" Before I could answer, we embraced each other for a long period of time, and for the first time, I was overcome with emotion as my eyes started to water with tears also. We were both silent; words were not necessary at this moment. I am not sure either of us could have talked right then anyways. Once we were both ready, we broke our embrace, and Joni said, "I almost stopped at Chandi's to tell her what had happened and ask her to come here with me. I didn't know she was in the plane with you, so I am glad I didn't. Ed was at the end of the service road when I turned in, and we stopped and talked briefly. Ed had told me everyone is safe, Pat, Chandi, and Steve. 'Chandi?' I asked. 'Chandi was in the plane too?'" Joni continued. "Steve, I was fine until I heard that. Where is she? I really need to see her!"

"I decided on the way here this morning I would invite her to go with us," I told Joni. "She has been asking to go for some time, and I thought this would be a good flight for her so she met me here. She is in the restaurant," I said while turning toward the restaurant doors. Joni was in a hurry to get there so I needed to follow her as she opened the doors and walked into the softly lit dining area. Chandi was still peering out the large window as Joni turned toward her and called out her name. "I am so glad you are all right," Joni said as she softly wept while they embraced. "I am thankful everyone is safe," she then stated.

"Please tell me how it all happened," Joni said to no one particular. Al came over again to offer Joni a seat, a soft drink, and some food. Pat opened the dining room doors and walked in and greeted Joni as they also hugged.

"Let's all sit down," I suggested as we gathered around the table. "Pat," I asked, "how is your wife and family?"

"They are fine," he replied. "They had not heard, so it was good they could hear my voice. It was really good, I mean really good to talk to them. I thank God we are sitting here together, alive and able to reflect on this day."

Pat, Chandi, Joni, and I sat around that table for some time sharing our thoughts and reflecting on each detail of our flight. We told Joni every minuscule detail of the flight, particularly the emergency approach into Bradford. I was able to reflect as if time had stood still for a period of time. It was good analysis we all needed. It was important therapy for each of us to share our inner feelings about the accident while it was still so very fresh. It provided Joni with much-needed detail and confirmation about what really happened in that airplane, the kind of detail she really needed to deal with the traumatic events of her day. It was important to cement the details of this day in our minds and for each of us to remind the other of the detail and sequence things happened in. As the pilot in command of the flight I had a need to revisit each decision I made, and each action I took for my own mental well-being. I felt responsible even though I knew the engine failure was not due to any action I had taken. I was confident I had done all I could to avoid the outcome, but needed confirmation by reviewing all decisions and actions with Pat, to see if I could have, or should have, done anything differently. I couldn't be sure I would ever pilot an airplane again. I was positive I needed to self-evaluate my skills. I was willing and ready to receive any criticism I deserved. What better person to do this for me than one of the flight instructors who taught me to fly, one that I respected personally and professionally, and the one who was in the airplane with me as things happened.

As the time passed and the daylight became darkness, we realized it was time to depart. Pat needed to drive back to Buffalo to be with his family, Chandi wanted to seek out some of her friends, and Joni and I wanted to visit my parents. And I still needed to call Rob, the owner of the airplane, to inform him.

We all walked out of the restaurant together and through the lobby doors. The cold, chilled January air felt good as Pat and I broke off walking toward his car. I was unsure what to say when Pat turned to me and said, "Steve, I want you to know I am proud of you. I saw you today under extreme pressure, and I wouldn't change one thing you did, not one decision you made in that airplane would I change. I want you to promise me something. I need you to promise me you will fly again. You need to. You are a very good pilot, and you can't let this incident stop you." Like wow, man, I am totally blown away. At this point, I am questioning my skills still wondering if I could have done anything to change the events of the day. Most airplane accidents are like a domino effect where one decision leads to another and so forth. If any one decision were to be changed, the outcome would be very different and the accident could be avoided. "Pat," I said, "thank you. But I cannot promise you that tonight. I need to think about this for some time. I need to talk to my wife, my kids, my parents, and my friends. I will need their support before I can commit to anything in the future. I can't have them worry and anxious every time I go flying, and even then, I just don't know. But you don't know how much that means to me. I needed to hear that from you tonight."

"I understand." Pat came back with "And I mean every word of it."

I shook Pat's hand and thanked him for the support. I then turned back toward my car and slowly walked over and climbed behind the driver wheel as Pat did the same. As I followed Joni in her car and Pat in his car around the circle leading to the service road exiting the airport, I took at deep breath and looked around. Chandi was with Joni, so for the first time, I was alone with my thoughts. I usually have the local country station on the radio while I travel, but tonight,

I needed quietness, so I reached over and turned the radio off. As I turned left toward Smethport, retracing the route I had driven to the airport that morning, I was finally able to allow myself to reflect on the events of the day and to enjoy the stillness of a silent trip to my parent's house.

THE NIGHT OF THE ACCIDENT

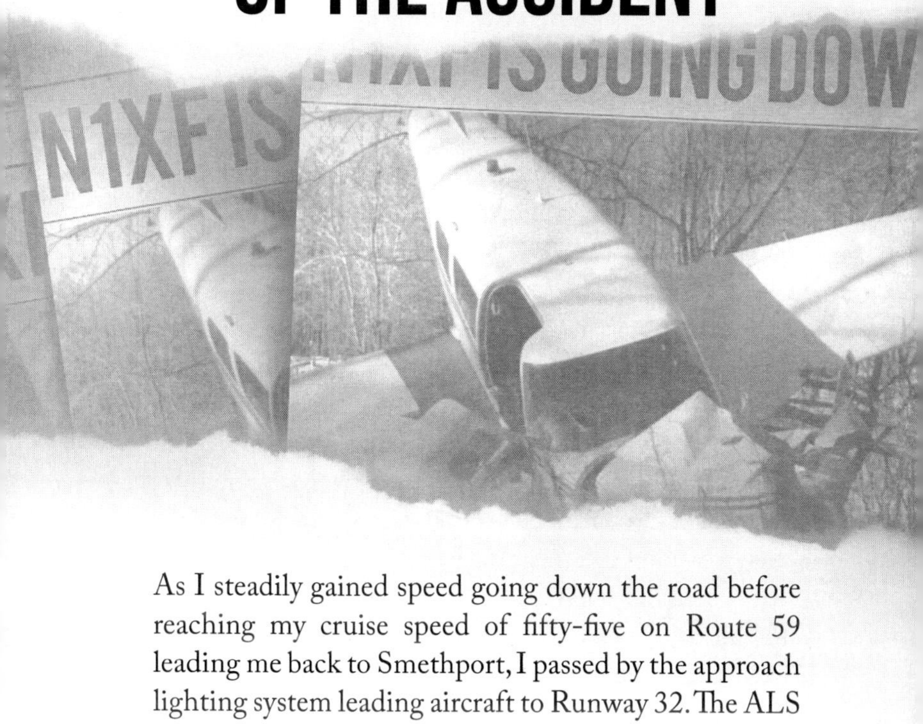

As I steadily gained speed going down the road before reaching my cruise speed of fifty-five on Route 59 leading me back to Smethport, I passed by the approach lighting system leading aircraft to Runway 32. The ALS is a lighting system installed on the approach end of an airport runway and consisting of a series of lightbars, strobe lights, or a combination of the two that extends outward from the runway end. ALS usually serves a runway that has an instrument approach procedure (IAP) associated with it and allows the pilot to visually identify the runway environment and align the aircraft with the runway upon arriving at a prescribed point on an approach. The lights were on indicating an approaching aircraft, and I instantly felt that I would be back in the air as a pilot in the near future. I felt the magical lure pilots feel, it was an immensely strong feeling, and I knew then I would fly again someday.

I knew the fact I was right next to the approach lights as they came on was more than a coincidence; I felt the presence of God with me and began to talk to God as I was crossing over the ridge and down the side of the hill into Smethport. I know he is all-powerful and has his ways of sending clear messages to us. I didn't question why the day had unfolded the way it did, but I just thanked him for being with us and keeping us safe. I strongly believe it wasn't my time yet, that I have much to accomplish in my life yet. I started to wonder what my purpose is here on earth, why some people die in accidents like ours and why we didn't. My thoughts wandered as I continued to pray without thought. It's a conversation God must have had trouble following; I don't remember what all I said, but he must have been laughing at me. It doesn't matter because it was therapy I needed, and he was providing the time and setting to make it happen.

I do remember being thankful that I didn't have to attend the funeral of my friend, Pat, or bury my beautiful daughter, Chandi. The thought of doing that was the mental plateau I had to break through. The thought that I could have been responsible for the death of my daughter was the most difficult thought and one I had to fight hard to overcome. I also remember feeling grateful my wife and my son Chad wouldn't be facing the remainder of their lives without their sister and daughter, father and husband.

I can only speculate that many other deep emotions and thoughts were with me during that time because, before I realized it, I was turning into the driveway at my mother and father's house. I don't remember driving through Smethport or the rest of the trip to their house or what I had been thinking of.

It was well into the evening by now, and as I pulled into the driveway, I suddenly realized that I had not called Rob to tell him what had happened with his airplane. Rob is a great guy and remarkably patient and understanding. I still wasn't sure what to say or what reaction to expect. My heart began to pound as I called out to Chandi and Joni as they were opening the door to my parents' house. "Hey," I hollered out, "I will be in soon. I need to call Rob and talk to him." I am sure the apprehension was apparent in my voice as they both turned and wished me luck.

I slowly flipped my phone open, searched my contact list, and put my finger over Rob's name as I hesitantly hit the speed dial. It seemed like an eternity as I listened to the small phone dial his number, and the phone began to ring on the other end. After only a couple of rings, I heard Rob's voice as he answered.

"Hi, Steve," he said as he answered. By the tone of his voice, I immediately knew he hadn't heard about the accident. My heart began to pound harder; I could feel the nervousness creep all over my body. "Rob," I said before pausing, searching for the right words to say. "Rob, I have had a terrible day today. On our way back into Bradford this afternoon, we lost complete power in the Comanche, and we crashed into the trees short of Runway Two-Three. Your airplane is totaled."

There was complete silence. In just a second or two, I continued, "My instructor, Pat, my daughter, Chandi, and I were on board and we all are fine. The three of us survived and no one is injured."

"Okay," Rob said. "No one is hurt?" he questioned.

"That's right," I said.

"Okay," he said. "I can replace the plane, no biggie," as he often said. "I will be in touch," he said and hung up.

I thought wow, that wasn't so bad and it went well, as I started to put my cell phone in my pocket. Almost immediately, the phone rang, and as I looked at the small display; I could see it was Rob calling me back. My heart started racing again as I answered it.

"Steve," Rob started with this. "Did you just tell me you crashed in the Comanche today?"

"Yes, I did, Rob," I replied as he cut me off.

"Did you say everyone survived? Did you say no one was hurt?"

"Yes, I did, Rob," when he again cut me off.

"I don't care about the airplane, Steve. I just need to know that everyone is all right. What happened?"

I quickly explained to Rob the events as they happened and he intently listened.

"Okay," he said again. "The only thing that matters is you and everyone else is safe. Don't worry about anything. I will talk to you when I get back in town."

"Thanks, Rob," I said. "And I am really sorry."

He gave me another "no biggie," and I hung up, glad I had that conversation behind me.

It was now time to see my mother and father. As I climbed out of my car and slammed the door behind me, I suddenly felt anxiousness about seeing my mom and dad, more so my mother. My mom is a small person, maybe just a little over four feet tall and weighing about ninety pounds. She's a sensitive person with the deepest faith of anyone I have ever met. This could be emotional, I thought as I opened the door to the entrance room.

When entering my parents' home, the living room is to the right of the main entrance door and is a long room. I walked in and turned and walked into the living room where my mom and dad were sitting talking to Chandi and Joni. Mom quickly stood from the couch and walked over to me and gave me a huge bear hug. Dad is one who shows less emotion and simply greeted me as he always does.

As I reached out to hug my mother, I distinctly noticed for the first time some extreme soreness in my body. My chest area started to feel sore and tight. I was positive it was the result of coming to a sudden halt after flying through the air at 105 miles per hour. There are two couches in the living room, and I sat down near the arm of the one nearest me. I looked over toward my mother, now back in her customary place on her couch and suddenly started to cry. A long emotional day had been kept bottled up inside me until this moment. Seeing the relief on Mom's face was enough to bring it all out of me. I was now able to share my emotion with my family in private. So much so that Dad became concerned and suggested I make a trip to the emergency room. He worried I would have dreams or nightmares about the accident and also felt due to the stiffness I was feeling, it would be smart to go to the Olean General Hospital emergency room for an examination. Joni quickly supported the idea. I insisted I was mentally and physically all right and just needed some time to vent my emotions.

After much prodding and encouragement, Joni and I were on our way to the hospital. I was becoming sorer as time passed and felt I would be more comfortable if I could be certain I was not injured. This was the first time we had alone together and gave Joni and me about thirty minutes to share our private thoughts.

Most emergency rooms on a weekend are extremely busy. As I walked into the emergency area, I was greeted at the desk by the receptionist at the triage area. "Good evening," she said. "May I help you?"

"Yes," I said as Joni stood by my side. "I was in an airplane crash this afternoon and would like to be examined." Her head shot up as she looked at me with in amazement and confirmed what I had just told her. I don't know how busy they were at this time, but the triage nurse was there almost instantly, and I was receiving priority care almost immediately, even as I assured everyone I was only stiff and sore and not experiencing additional pain.

I once again began explaining the day's events to the triage nurse in detail as he intently listened to my story. All accidents have the potential to be serious, but aircraft accidents have a mystic to them that gathers the attention and peaks the interest of most people, even those in the medical field. It was evident to me I would quickly be examined by the medical staff simply based on the type of accident I had been involved in. "I cannot believe you walked in here" was a common comment I heard in the emergency room.

The emergency room physician was extremely meticulous as he ordered all the pertinent X-rays and tests. He asked me a wide array of questions as he determined I didn't have any injuries. He assured me I was only experiencing the normal soreness of a traumatic event. He offered me the normal medications for such soreness, which I refused, confident I could physically and mentally handle the recovery period.

As I lay there on the emergency room bed, my thoughts started to turn toward the investigation stage that I knew would follow. I began to anticipate what the investigators could be considering as they looked at every detail of our flight. My defensive mechanism started to kick in, and I decided I wanted to be administered a drug and alcohol test. I reasoned that would be one area the NTSB would look at, and I felt the need to eliminate any possible accusation in regards to alcohol and drugs. This idea was met with resistance from the medical staff and the doctor on duty. I continued to insist this test be ordered, and they persisted to say no. Finally, the doctor came to discuss it with me. He seemed perplexed by my continued insistence until I explained my reasoning. "I understand your concern," the doctor assured me, "but my concern is the investigators will look at this test and question why I ordered it. Normally we administer drug and alcohol tests when we have adequate suspicions to do so. I am not seeing any indications that would lead me to believe such a test is necessary… Subsequently, you should understand such a test might create a problem for you instead of preventing one."

I understood what he was telling me but still wished to make sure I couldn't be falsely accused of being under the influence of any substance while piloting an aircraft. After some more discussion, the doctor suggested we both sign an affidavit that stated a test was ordered and administrated strictly due to my insistence; also that he hadn't observed a reason to justify the test and only agreed to it due to my persistence.

I cannot be positive this had any effect on the investigation, but I wasn't questioned about drug or alcohol abuse. It would have been a waste of their time anyways, but the test was ordered, and I had the security I wanted, even if it might have been superfluous.

Joni and I walked out of the hospital with the good wishes of the emergency medical staff and the great fortune of our lives yet to live and ready for the media and investigation stage.

THE DAY AFTER

Joni and I had been planning to go to Erie, Pennsylvania, the next day, a Sunday, with a couple we were friends with. Our friends lived in Little Valley, New York, so we had arranged to meet at the McDonald's in Salamanca, New York, and we would car pool to Erie from there. On our short trip from the Olean General Hospital on Saturday evening, Joni and I briefly discussed canceling our trip but decided not to. There were two main reasons for this decision; the first was we were going to attend a show in the theatre district of Erie, which we were very much looking forward to, and the second was I really felt I wanted to stay active and busy the day after.

We both slept good on Saturday evening and were well rested for the new day as the alarm sounded for us to get around for the trip to Erie. As I rolled over onto my side and pushed myself up from my bed, I noticed the bright sun shining through our bedroom window. I walked over and stared out much like I had done the previous day. I was struck by the resemblance of this day in comparison to yesterday. The sun was bright, the sky a beautiful blue, and a light wind from the northwest was again swirling the loose snow around our small lawn and driveway. My thoughts drifted back to the morning before when I was anxiously preparing for our flight to Syracuse. As I stood there peering out the window, I was thinking, "What a difference a day makes!"

We hurried around as we readied ourselves for our trip to Erie. I could tell the temperature was still hovering around zero as I could see the small ice crystals in the air and glistening on the car surfaces outside. I was ready to leave before Joni and hurried out to the car to start it and let it warm up. I parked the car in the driveway and waited in it for Joni to come out to get in. I had parked at just the right angle where I could see customers entering and exiting the main entrance doors to the Country Fair store. As I sat there watching the people, I noticed several customers leaving with the *Olean Times Herald* newspaper in their hands. This wasn't at all unusual since the Sunday paper was always a huge seller on Sundays.

But on this day, I knew the news of our airplane accident would be on the front page. Air mishaps routinely are front page news regardless where they happen, but they almost always are in our small rural area. I wondered how many people would read about our crash and how many of those people know who I am. My stomach started to turn as I could feel the emotions churning inside me, indicating I wasn't ready to read the article or face what the press might be writing. I was once again deep in my thoughts because Joni was already in the car, and I hadn't heard her get in.

Our trip from Olean to Salamanca where we were to meet our friends is about thirty minutes traveling west on Interstate 86. We quickly exited our driveway and found our way through the city to the ramp leading us to Interstate 86. Soon we were traveling west and arriving at our destination in Salamanca. As we turned right into the parking lot at McDonald's we saw Jim and Mary's vehicle parked in the lot waiting for us. We were a few minutes early, but they are always prompt and usually arrive several minutes early. Mary was already in the back seat, and Jim had planned to drive so he was waiting behind the driver's wheel. As we usually did, Joni opened the door and slid into the right side rear seat, and I got in and sit on the passenger side front seat.

As I was buckling my seatbelt and securing my shoulder strap, Joni and Mary were going through their "good morning, how are you?" routine as Jim and I exchanged our greetings. Mary held up the *Olean Times Herald* and said, "What is this all about, Steve?" Mary was not aware the oversized picture above the fold on this Sunday's *Times Herald* was that of the plane I was piloting or even that I had been involved in an aviation accident. I turned to look at the newspaper and started to weep. The headline read, "Three Walk Away from Crash" and was the feature story on this Sunday morning. With it was a big picture of the totaled airplane laying harmlessly in the woods with the door wide open.

The look on Jim and Mary's face was one of total surprise since, at that time, Pat, Chandi, and I had not been identified by airport management, the FAA, or the NTSB. In this small area, I was sure it wasn't hard to figure out for those with an ear and eye for the local aviation community. The media was only aware that a male flight instructor was piloting the airplane and was accompanied with a second male flight instructor as well as an adult female. I do not know the reason everyone was being so tight-lipped about our identity, but it was certainly an approach I could endorse for a short period of time.

I didn't want the media calling my house or cell phone for a comment until I was able to discuss the accident with legal representation. I was already becoming greatly aware that all my actions and decisions would be scrutinized by the investigators. I believed what had happened was not avoidable given we were in the airplane, and we had experienced mechanical failure beyond our control. I was also comfortable, based on my training, that Pat and I had systematically and properly executed the emergency procedures as we were trained.

Equally, I was confident I had followed the emergency procedures for the Piper Comanche as described and provided in the aircrafts handbook and on the checklist provided with the airplane.

"Oh, I am so sorry," Mary said as she put her hand on my shoulder. "I honestly had no idea and only thought you would be able to give us the scoop."

"We honestly didn't know," Jim chimed in.

"I am sorry," I said, "but seeing the picture and headline is traumatic for me."

"What happened?" Mary asked and continued by saying, "You don't need to tell me if you don't want to."

"It's okay," I replied. "It will probably be good for me to tell you the story."

"The whole day and all the details are truly amazing," Joni offered. "Just listen to how it all happened, how I was notified, and how I found out Chandi was involved. You won't believe it! God certainly was involved and kept them safe."

By now, we were leaving the McDonald's parking lot and ready to get back on Interstate 86 for our westbound trip to Erie. It was approximately an hour and fifteen-minute drive to Erie and most of that time was necessary for Joni and me to share the details of the day before with them. One of the most common questions I still get when discussing this accident with others is "How scared were you?"

To answer that question accurately and honestly without sounding macho or disconnected, I need to explain my thoughts. But the short answer is this: amazingly, not very scared. I know that doesn't sound true, but it truly is. Let me explain the reasons as I view them:

1. First I was busy and didn't have time to think about what could happen. At the first sign of trouble, my training took over, and I immediately reached for the emergency checklist. Pat instantly was reminding me to get to the best glide speed. My total concentration was on executing the checklist and solving the failure. I was able to do this because I knew Pat was a very good pilot, and I delegated the responsibility of the radios and piloting to Pat while I troubleshot.

2. By the time I had exhausted all possible remedies available to us, we were just seconds from going down. I believe if I had more time, I might have reacted differently. Even as we started to clip the trees, we were busy instructing each other to be sure we were prepared for impact. As odd as it may sound, the timing was perfect for us.

3. The thought I have the most is "Why didn't the thought of dying in N1XF ever seriously enter my mind?" I cannot give absolute answers because I still ask myself that regularly. What I do know is I felt a confidence and calmness that cannot humanly be explained. Except of course but for the presence of God in the airplane with us. Those who know me will generally confirm I am not macho, I am conservative, and I don't take unnecessary risks. I totally recognize the

inherent dangers of flying, but I must add, I strongly defend the safety of flight. I know I could die in an airplane, I just knew I wasn't going to die in N1XF, and I had a strong feeling we all were going to walk away and live to share our experience. So much so that at the instant of impact, I made a conscious decision to keep my eyes open. Maybe I am stubborn, maybe I am defiant and wasn't going to give in that day, and after all, I do hate to lose. But I did keep my eyes open, and I have a distinct visual memory of the wings hitting the trees, of the Plexiglas windshield breaking into seemingly millions of tiny pieces and flying all around us. I can still visualize N1XF starting to spin to the right and can still clearly see the nose hitting the ground and bending to the left.

4. Finally, I am not afraid of death. I believe in John 3:16: For God so loved the world that he gave his one and only Son, that whoever believes in him shall not perish but have eternal life. God promises us a better place than life on earth, so what is there to be afraid of? Of course I am not ready to leave yet either, so I didn't hope or wish for death then or now. I also believe when God is ready for me, he will take me and I will have little or no choice in the matter. That is the belief that has helped me log hundreds if not a thousand plus flights since the accident. I simply knew God was giving me peace and comfort in that airplane in that moment and it is the only

> reasonable explanation. My goodness, I didn't even feel the need to pray at the time. Like I mentioned, I am stubborn!

So now you have the short and long reason I believe I wasn't really scared. I think that also contributed to my daughter remaining calm although she didn't forget to pray. Maybe she was praying so hard that it was enough for both of us!

But back to our trip to Erie, it was a great day and kept me involved in an activity that made our day wonderful and memorable. As we returned home that Sunday evening, I was thankful Joni and I had decided to go to the theatre as planned. It was the relaxing day we both needed and helped prepare us for the week ahead.

THE MEDIA AND THE INVESTIGATION

I had enjoyed great support from the local media over the couple of years I was operating as a flight instructor at the Bradford Regional Airport. I didn't want to disrespect their need to report this accident accurately, and as of Monday, they were following up the accident with coverage that our names were still not being released to the public. We didn't have anything to hide from investigators or the public, I personally didn't, and I myself didn't understand why the apparent secrecy. It was somewhat uncomfortable, and I was torn between making myself available to the media or remaining silent as I was advised to by the legal council I had through the Aircraft Owners and Pilots Association. I had purchased legal representation for such instances when I joined the AOPA years before; therefore, felt it was necessary I trusted their advice. I chose to follow the advice of my legal council and anxiously remained unnamed, knowing the media probably was well aware I was the pilot in command. I felt like a "sneak" and knew it was only a matter of days until our identities would be publicly known. My other concern was the pressure Rob would be under to identify us. It was accurately being reported he was the owner of N1XF, and the longer time passed, the more skeptical the media would become toward him. Rob is a trusted major businessman in our area who enjoyed important friendships and trust within the local medias, and I didn't want to compromise his position in the community, especially when there wasn't any good reason to do so.

Rob and I made time on Monday to discuss the accident and to provide him every detail of the flight and to provide answers to all of his questions. He grilled me on all the preflight actions and about how decisions were made on board the airplane. We discussed the importance of total cooperation with any official investigator and all authorized investigating agencies as well as the desire for transparency when supported by legal council.

I am positive Rob was concerned about potential civil lawsuits, but it wasn't ever discussed. He always outwardly projected his concern for the safety and health of Chandi, Pat, and I; and I know he was sincere. In turn, I wasn't going to ever consider civil suits regarding this accident and hoped Pat and Chandi would concur; although I couldn't be positive as the subject hasn't ever been discussed among us. That is not part of this story, but I think is important to mention for the reason that it is important everyone understands decisions were not made to further the possibility of civil suits or to defend against them.

Federal regulations require operators to notify the NTSB immediately of aviation accidents and certain incidents. An accident is defined as an occurrence associated with the operation of an aircraft that takes place between the time any person boards the aircraft with the intention of flight and all such persons have disembarked, and in which any person suffers death or serious injury, or in which the aircraft receives substantial damage. An incident is an occurrence other than an accident that affects or could affect the safety of operations. The NTSB had already been notified by Pat on Saturday afternoon while we were being examined in the fuel office. On Monday Form 6120.1: Pilot/Operator Aircraft Accident/Incident Report needed to be filled out and faxed to the NTSB. This form supplies all the information needed by the NTSB to completely investigate accidents. Nearly every piece of information in regards to the aircraft and its maintenance is supplied on this form. In addition, we needed to supply great detail about the aircraft owner, the operator, and what part of the federal aviation regulations we were operating under, which was Part 91. We also needed to detail the mechanical malfunction that lead to the accident and the purpose of the flight. Airport information such as runways available, condition of runways, and other conditions were required. Other information we had to supply was the entire flight itinerary plus the fuel services available at the airport along with the services we used prior to departure. The report went on to request information about the weather the day of the accident and particularly at the time of the accident. Also, we needed to supply to the NTSB on this form the complete pilot history of both Pat and me. Information like total flight time, ratings, and endorsements and time in the type of aircraft we were flying. Also, we needed to supply information about our passenger, Chandi. Lastly, we were required to supply a complete narrative in chronological order about the flight with emphasis on the circumstances leading to and the nature of the accident. We were required to state the time and point of departure, our intended destination, and any services we obtained before and during the flight. Finally, we were asked to describe the terrain and wreckage distribution the best we could.

As one could imagine, this report was intensive to fill out due to the enormous amount of information required, and because we needed to be totally accurate with every detail. This would be our official statement, so accuracy of all information was imperative. Gathering the information and verifying its accuracy consumed most of this Monday.

Also on Monday, the local Flight Services District Office (FSDO) out of Pittsburgh contacted Pat and I as part of their responsibilities. The Pittsburgh area FSDO was the local office with jurisdiction over all operations around the Bradford Airport. Pat and I were both asked to supply a statement of accident to the local FSDO office. Following is my statement as faxed to the FAA regarding this accident, minus Pat's last name and phone numbers:

> This is my statement of events regarding the accident at Bradford Regional Airport on Saturday, January 10, 2004.
>
> I had asked Pat, CFII, to come to Bradford and fly with me to sharpen my instrument skills. This was the intention of the flight. We decided to fly to Syracuse, NY for approaches and holds then return to Bradford Pa. I also invited my daughter to join us for the flight since it was a beautiful day to fly.
>
> Before departing BFD, I requested that the main tanks be topped off which was done as we were standing there. In addition, the preflight was completed in a through manner. We departed BFD just before noon. I do not recall the exact time but remember our clearance being void at 1700 GMT or approximately then. The flight to Syracuse was uneventful. After completing our last approach to SYR, I was given a heading I believe of 280 degrees and cleared to 6000, our cruising

altitude. At about 4,000 feet, or maybe 4,500, we experienced something that was not normal. The plane seemed to miss or as Pat referred to it at the time, hiccupped. It definitely got both of our attention as we both stopped what we were doing and checked all instruments for any indication of trouble. All instruments indicated normal, so we decided to proceed and note it on the ground in BFD. We obtained cruising altitude and everything seemed normal.

Approximately 6-8 miles Northeast of the BFD VOR I was given clearance to descend to 4,000. Our intent was to fly the VOR 14 full procedure and circle to land on 32. I believe the time to have been about 1900 GMT as I looked at my watch when I started the decent and recall this approximate time.

As I started my decent, the engine suddenly lost power. We experienced several minor power surges and then it quit. I told Pat to take the airplane while I went through the cockpit checklist and attempted to restart. I promptly cycled through all tanks trying to find one that would help us. Pat watched me do this while I talked myself through it. I was obviously unsuccessful at the restart.

We obtained best glide speed and initially went for Runway 32 but decided together that 23 offered the best opportunity. We originally thought we could make the runway, but then hoped for the clearing just before the runway. Pat fought the tendency to pull back and flew the airplane level. The last thing I remember is looking out and thinking, we are at best glide,

> we are level and the trees are small and that maybe we have a chance to survive.
>
> After we spun around to a stop, I looked over to make sure Pat was conscious, and he was and I then verified that my daughter was all right Pat opened the door and exited the aircraft, I slid out and Pat dragged my daughter out. When we were all on the ground I immediately instructed everyone to get away from the aircraft.
>
> To the best of my knowledge, these are the events that preceded the crash, in the best order I can recollect. If you should have any further questions, please feel free to contact me at xxx-xxx-xxxx or on my cell at xxx-xxx-xxxx.

Pat submitted his statement of accident at the same time, and I assume it was supportive of and consistent with my statement since everything I have written or stated is totally true. I also believe since I don't remember any other contact from the FSDO or any agency that Pat's recollection was in conformity to mine, as it should have been.

On Tuesday, I received a phone call from Kevin who is an insurance claim specialist. Kevin was reaching out to Pat and I, concerned about our physical condition since it had been about seventy-two hours since the accident. Sure Kevin was concerned about financial responsibilities of the insurance company, which insured the aircraft and owner; however he didn't apply any pressure to make a statement of well-being. In fact, several times during our conversation, Kevin encouraged me to seek any medical advice or examinations I felt necessary and to be absolutely sure I didn't have any lingering problems caused by the accident. About two weeks later, Pat and I received a non-injury statement Kevin requested we fill out and return to him, which we did. At the same time, Kevin brought to our attention a portion of the policy that would cover any out-of-pocket medical assistance we might need in the future due to the accident.

By Tuesday afternoon, the *Bradford Era* was starting to become anxious for names of the aircrafts occupants. I received a voicemail message from a reporter following up on the accident, wondering if I could provide any information about the accident since all other sources were seemingly not cooperating. I knew it was getting to be time to make a statement but felt the need to handle all my legal responsibilities first then to document the details so I wouldn't be misquoted or that any comment could be taken out of context. Each day, a small article would appear in the *Bradford Era* stating that we had not been identified yet. I again elected not to return the reporter's call before I could touch base with my legal council.

It was midmorning on Wednesday when I received a call from Rob. The *Bradford Era* management had contacted Rob asking for information. As Rob relayed their conversation to me, they knew who was piloting, felt they knew who the co-pilot was, and weren't sure who the adult female was. In respect for Rob, they had not revealed our names but couldn't wait much longer.

"Can't we make a statement about who was on board and what happened?" Rob asked.

"We don't have anything to hide, but it certainly is starting to appear that way," he concluded.

"I agree," I replied. "Just let me call the attorney through the AOPA Legal Service and discuss it with him."

"OK, get back to me ASAP," Rob asked of me, and I agreed.

I immediately called my contact with the legal service I was using and was able to discuss the situation with an attorney. He sort of chuckled and stated, "Those people up there need to get a life."

I simply said, "This is a small rural area and this is big news here. I need to be as transparent as possible with the media. They have always been supportive of Rob and I, and I owe it to them if possible."

"Then tell them what they already know" was the attorney's response.

"Thank you," I said and decided to issue a press release immediately. I called Rob back to inform him I would type a press release within minutes and distribute it to all local papers and radio stations. I prepared the press release and faxed it to all local media. The press release was as follows:

> To: *Bradford Era* News Room
> From: Steve Jackson
> Date: January 14, 2004
> Re: Accident on Saturday January 10, 2004

> In response to the voice message I received yesterday afternoon, I do confirm that I was in the aircraft that crashed at Bradford Regional Airport on Saturday afternoon. Also in the aircraft was Patrick (last name deleted), a CFII and my daughter Chandi Jo Jackson.
>
> We experienced engine trouble on our decent into Bradford Regional Airport and had to execute an emergency landing. There are no injuries and the three of us walked toward the airport until we secured assistance.
>
> For any additional information, I refer you to the FAA or NTSB.
>
> Sincerely,
> Steven B. Jackson
> Cc: Olean Times Herald

Well, that was enough information to make front page headline the next day. This information was enough to confirm what needed to be publicly revealed and the media pressure was off. I am glad I did it when I did and believe the timing was right.

My main occupation is that of an income tax preparer, so I was getting extremely busy at the office. Probably a good thing for me since I really wanted to keep my distance from the investigation and had heard that N1XF was in pieces in one of the hangars at the airport as the NTSB was conducting their investigation.

Surprisingly to me, I heard nothing from any investigator and wasn't asked any further questions. The only correspondence was a request by me to recover Pat's logbook. I was told where I could pick it up at and did so then met Pat at Burger King in Ellicottville, New York, to return it to him. That was it! I wasn't informed when a determination was made or published about the cause of the accident. I only assumed that if I were found at fault, I would be notified or required to attain additional training before I could resume my flying or instructing career. I did make a trip to the airport, and under the supervision of some airport personnel, I viewed the wreckage in the hangar as Chandi was preparing a short article for a publication related to the college she worked for. I found out about the final determination one day through a phone call from Chandi who regularly monitored the progress as it was posted.

"Dad," Chandi said, "the final report about our crash is online." My heart sank! I felt both relief and anxiety as I went to the NTSB database to finally read what the official determination of cause was.

Following is the complete narrative of the final determination as posted on the NTSB website:

> NYC04LA057
>
> On January 10, 2004, about 1425 eastern standard time, a Piper PA-24-260, N1XF, was substantially damaged during a forced landing, following a total loss of engine power while on approach to Bradford Regional Airport (BFD), Bradford, Pennsylvania. The two certificated flight instructors and a passenger were not injured. Visual meteorological conditions prevailed for the flight that departed Syracuse Hancock International Airport (SYR),

Syracuse, New York. An instrument flight rules flight plan was filed for the personal flight conducted under 14 CFR Part 91. The pilot-in-command (PIC) in the left seat stated that he planned to practice some instrument approaches at SYR, with the pilot in the right seat acting as a safety pilot. Prior to departing BFD, the main fuel tanks were "topped off." The PIC further stated that the airplane had a supplemental type certificate for two fuel selectors; one for each wing. Each fuel selector could be positioned to "Main," "Aux," "Tip," or "Off." The flight departed BFD about 1200, with both fuel selectors positioned to "Main," and flew uneventfully to the SYR area. The PIC completed several instrument approaches, and began a return to BFD about 1330.

Approaching the BFD area, the flight was cleared to descend from 6,000 feet to 4,000 feet, and the PIC intended on flying the VOR RWY 14 approach. During that time, while both fuel selectors remained positioned to "Main," the engine "hiccupped." About 6-7 miles northeast of the BFD VOR, while descending, the engine experienced several minor power surges, and then lost all power. The PIC established a proper glide speed, and the safety pilot initially suggested a right base leg for runway 32. However, the PIC chose runway 23 because it was a straight-in approach from his position. During the glide, the PIC completed the emergency checklist items, which included verification that the electric fuel pump was on, and cycling the fuel selectors through all tanks. The PIC was unable to restart the engine, and continued the glide toward runway 23. The PIC maintained glide speed, and initially thought he would reach the runway, but the airplane impacted small trees about 2,500 to 3,000 feet prior to the runway. The PIC added that there were no suitable fields near the airport.

Examination of the engine by a Federal Aviation Administration (FAA) inspector did not reveal any pre-impact mechanical malfunctions. The inspector observed fuel in all fuel tanks, and fuel was present from the selector valve, to the fuel servo, to the flow divider. When power was applied to the electric fuel pump, it operated continuously. The inspector noted that the right fuel selector handle was difficult to move, with some debris lodged in it. He also observed that the throttle idle adjustment bushing was missing. However, the inspector could not determine if the bushing separated during flight, or during the forced landing. Subsequently, a flow test of the fuel servo and flow divider was conducted at the manufacturer's facility, under the supervision of another FAA inspector. Except for the number six port on the flow divider, which flowed 8 percent lower than the minimum allowed value, the units tested within limits.

From the query page;;

Prior to departing his home airport, the pilot topped off the main fuel tanks. The flight departed with both fuel selectors positioned to

"Main." After an approximate 2 hour flight, the pilot was approaching his home airport, and the fuel selectors remained on "Main." While descending from 6,000 feet to 4,000 feet, the engine "hiccupped," and then experienced several minor power surges, before losing total power. The pilot established a proper glide speed, and attempted a straight-in approach for the nearest runway at the home airport. During the glide, the pilot completed the emergency checklist items, which included verification that the electric fuel pump was on, and cycling the fuel selectors through all tanks. The pilot was unable to restart the engine, and continued the glide toward runway 23. The pilot maintained glide speed, but the airplane impacted small trees about 2,500 to 3,000 feet prior to the runway. Examination of the engine did not reveal any pre-impact mechanical malfunctions. Fuel was observed in all fuel tanks, and fuel was present from the selector valve, to the fuel servo, to the flow divider. When power was applied to the electric fuel pump, it operated continuously. Further examination of the engine revealed that the throttle idle adjustment bushing was missing. However, it was unknown if the bushing separated during flight, or during the forced landing. Subsequently, a flow test of the fuel servo and flow divider did not reveal any discrepancies that would have resulted in a total loss of engine power.

The National Transportation Safety Board determines the probable cause(s) of this accident to be:

> A total loss of engine power for undetermined reasons during approach.

These two above narratives can be found on the NTSB website at the following URLs;

> The query page:
>
> http://www.ntsb.gov/aviationquery/brief.aspx?ev_id=20040122X00086&key=1
>
> Full narrative:
>
> http://www.ntsb.gov/aviationquery/brief2.aspx?ev_id=20040122X00086&ntsbno=NYC04LA057&akey=1

So the investigation is over, and I am back in the air! Everything I have written about to this point is based on fact and what I recall to the best of my ability. I have referred to a file full of notes to refresh the details in my mind and the conversations as I recall them. It amazes me the detail I remember. I hope the writing here not only serves the purpose of accurately recording the events of January 10, 2004, as they pertain to my day and experience but also helps others to understand what it is like to live through such an experience.

The next and final chapter moves from discussing facts to personal observations, thoughts, and feelings as they pertain to aviation and, in particular, "N1XF going down."

FINAL THOUGHTS

This is the most difficult chapter to write. There have been many thoughts and events since the accident that have changed my life. In addition, I have had several years to think about what happened on January 10, 2004. Today I have great recollection of the events of that day, which the story as I have written it should support. That cold January day and what we experienced during that time in our lives has never been far from my mind. In life, I tend to look forward and not back; I believe I have succeeded in doing so with this accident also. I returned to flying within weeks of the accident with the full unconditional support of all my family and friends. The first opportunity I had to leave work early when the weather was cooperative, and my cousin Ed was available to fly with me, I climbed back into an airplane and made the difficult first flight.

Although I have moved on from the accident, the mere fact I chose to write about it some nine years later also is an indication I still need to put my thoughts in writing. That I probably still have issues that must be addressed, so this chapter is dedicated to outwardly expressing those inner thoughts and feelings that still exist. These are questions I will never have the answer to. There were circumstances and sequences of events that day that can only be explained by faith and my belief that God truly was involved and in complete control of N1XF.

One of the hardest emotions to overcome was the knowledge that I had the life of my daughter in my hands that day, and I nearly flew her to her death. Since the accident, Chandi has met her husband, Matt, and has become a brilliant stepmother to Hadden. Hadden is a truly superb and awesome young man and now a teenager. I cannot imagine life without Hadden a part of it. He is also a loving role model big brother now to Broden and Landen. Broden is now five and full of love and life. Very often as Broden and I wrestle on our living room floor, my thoughts flash back to the day of the accident. Before our wrestling bout is over and he has won, I usually take the opportunity to look into his beautiful big blue eyes, deep into his soul and tell him I love him. I cannot resist thinking how close we came to a different ending, and these boys wouldn't have been possible. It serves exceptionally well as a reminder how God is in control.Broden is such a warm and beautiful person! And Landen is only one and starting to build that relationship and bond with me. I look at these three boys and start to realize God didn't need a grandstand reason to allow us to survive the accident. Or maybe he did and they are it. I will just allow myself to love and enjoy them.

The events of the day still astound me. I never wear gloves and hadn't purchased a pair in years—hadn't worn gloves in years. Why did I impulsively stop at Costa's Hardware that morning with the intent of purchasing gloves? Gloves that I would need a few hours later as we trudged through the woods looking for help; gloves that seemingly disappeared and haven't been located since I took them off, whenever and wherever that was.

How is that I rarely had cell phone service at the airport and almost never in the fuel office, but on this day, cell service was strong enough that Joni could hear everything as it was unfolding in the fuel office that afternoon? Why didn't I hang the phone up when I fully intended to and thought I did? And why was the timing so perfect that I couldn't personally talk to her, she only needed to listen?

And how is it that I decided to invite Chandi to fly with me on this day? We have always had a good relationship, and now we enjoy a bond few father and daughters do. We have a date and time that we always communicate with each other. It is special!

And most times, the engine is pushed back into the fuselage, killing the occupants, but not this day. A tree made sure of that, and it also stopped us from flying into the pond a few short yards ahead of us. And did I mention we came to rest close to a logging road, one where an operator was busy working? Sure I did, but how did this happen and why did it happen? There are just too many domino effect miracle pieces to this day to ignore them all.

I have experienced so much in life since the accident and met many wonderful new friends. I've trained several new pilots and helped to form a flight club at the Bradford Regional Airport. It's an organization that all members are proud of and rightfully so. We are furthering general aviation in our area and doing it successfully. We did it the right way and like to think we are a role model for the future of general aviation and aviation clubs. I encourage you to check us out at http://twintieraviation.com.

This brings me full circle to the beginning of the story. On a regular basis, I am still stopped on the street of my town and asked about the accident. Time and again, people I meet for the first time ask me about the accident. I have a story to share, and they are interested. I also give God credit for this opportunity, and I hope that testimony is transparent also.

N1XF did go down on January 10, 2004. Pat, Chandi, and I walked away on foot grateful to be alive.

We're alive. Thank God we're alive!